Six Sigma

Timothy D. Blackburn

Six Sigma

A Case Study Approach Using Minitab®

 Springer

Timothy D. Blackburn
The George Washington University
Washington, DC, USA
Datovation@gmail.com

ISBN 978-3-030-96215-9   ISBN 978-3-030-96213-5   (eBook)
https://doi.org/10.1007/978-3-030-96213-5

This Springer imprint is published by the registered company Springer Nature Switzerland AG
The registered company address is: Gewerbestrasse 11, 6330 Cham, Switzerland

Preface

Six Sigma has been a topic of great interest for many years and remains so but can be elusive for many. Six Sigma is a problem-solving technique that leverages DMAIC phases (Define, Measure, Analyze, Improve, and Control) and a data-centric approach, leveraging applied statistics. It seeks to solve novel problems where there isn't an apparent root cause, or a solution identified.

Achieving certification and becoming proficient at Six Sigma can take years and multiple projects to achieve proficiency (at least for robust programs of instruction, coaching, and certification). This text is intended to walk the reader through a (imaginary, fictional) case study, explaining both the approach and how to use Minitab® in a practical way, with the belief that this approach will expedite learning, and enable the reader to quickly apply tools and methods. Also, it will be a helpful resource and reference for tools and methods.

The imaginary company used in this book is KIND Karz, a new type of automobile company that started off successfully but recently has experienced increasing recalls and warrantee claims. These have further led to reduction in sales, and a falling stock price. The book will guide the reader through the phases of DMAIC to arrive at root causes and corrective solutions. Included are data sets and step-by-step instructions on how to analyze data in the context of Six Sigma using Minitab®.

The book is structured to cover each DMAIC phase, to a level of comprehension expected of a Six Sigma team member. Then, additional sections are added within the Measure and Analyze phase chapters for the Six Sigma practitioner or project leader, where the step-by-step instructions are provided for the statistical tools covered in the introductory DMAIC section. Other material is also covered, including detail of the case study, typical storyboard contents and examples for the case study, and datasets.

The techniques included in this book are expected to be between the Green and Black Belt certification levels and represent some of the most common tools and methods used in practice based on the author's experience. This book is suitable as a text for Six Sigma or Minitab®/statistical methods at the college level, for Belt certification training, or as a self-study guide and reference.

The KIND Karz case study is fictional, with fictional products, situations, root causes, and technologies, and includes generated data for purposes of illustrating the Six Sigma tools and methods. No inferences are to be made to other products, data, equipment, or processes. Also, any reference to companies or people is purely coincidental and unintentional. Further, no warrantee is provided or implied through the application of techniques described herein.

Minitab® versions 18, 20, and 21 were used in the analysis and for images in this book. Also, versions 18–21 were confirmed for compatibility with the content included. The book assumes general familiarity with Minitab®, but for the new user refer to Appendix A.

Any use of the KIND Karz case study and data other than individual learning, including but not necessarily limited to training or teaching others, is not permitted without permission of or license from Timothy D. Blackburn © 2020, 2021, 2022.

Washington, DC, USA   Timothy D. Blackburn

Acknowledgments

I could not have pursued my academic and many work interests to a level that would have enabled authoring this book without my wife Ana. She has done so much to encourage me and has selflessly managed matters that otherwise would have distracted. And my children have also been supportive, tolerant, and proud of their dad (and I am even more proud of them) — thanks Chris, Albert, and Bekah. Also, a shout out to my parents, Marvin and Betty. They have been a continued source of encouragement for whatever next thing has captured my attention.

Thanks also to my students, and the continual questioning that increases my learning as well. Also, thanks to The George Washington University, where for many years I have been permitted to explore my curiosity, both as a student and faculty.

A great big *thank you* goes to the reviewers who provided their expertise and considerable time to review and comment on the manuscript, especially Elie Talej and Isa Bin Amir Hamzah.

Finally, thanks to God who instilled in me a desire to understand maths, the language of substance and the truth of things.

Contents

Acronymns

ANOVA	Analysis of Variance—hypothesis test for three or more sample means
CI	Confidence Interval
Cpk	Process capability (short term or within batch)
CTQ	Critical to Quality
DMAIC	Define, Measure, Analyze, Improve, and Control
DPMO	Defects per Million Opportunities
FMEA	Failure Mode and Effect Analysis
HVAC	Heating, Ventilation, and Air Conditioning
Hurdle Rate	Desired rate of return
IPO	Initial Public Offering
IRR	Internal Rate of Return
LCL	Lower Control Limit
LSL	Lower Specification Limit
MR	Moving Range
MSA	Measurement System Analysis
NPV	Net Present Value
OpEx	Operational Excellence—a common name for a function responsible for continuous improvement and Six Sigma
Ppk	Process performance, or long-term process capability
R&R	Repeatability and Reproducibility
SIPOC	Supplier, Inputs, Process, Outputs, and Customer
SME	Subject Matter Expert
TPS	Toyota Production System
UCL	Upper Control Limit
USL	Upper Specification Limit
VIF	Variance Inflation Factor
VOC	Voice of the Customer

About the Author

Timothy D. Blackburn, is a summa cum laude graduate from the William States Lee College of Engineering (UNC-Charlotte) and holds an MBA from the Kenan-Flagler School of Business (UNC-Chapel Hill). He also received a PhD in systems engineering from The George Washington University **(GWU).**

He is a licensed Professional Engineer in multiple states and holds a Master Blackbelt in Six Sigma. Currently, he is Professorial Lecturer in Engineering Management and Systems Engineering (EMSE) at The George Washington University where he teaches multiple master's and doctoral level courses and serves as research advisor to doctoral students. He is frequently a committee member for dissertation and praxis defenses. Tim is widely published in peer-reviewed journals and has spoken at professional conferences. He also works in the pharmaceutical industry providing internal consulting in many of the areas covered in this book. Contents herein benefit from over 30 years of experience in both industry and academia.

Tim's affiliations (current or previous) include the Tau Alpha Pi and Phi Kappa Phi Honor Societies for academic achievement, the International Society for Pharmaceutical Engineers (ISPE), American Society for Quality (ASQ), Pharma Engineering Roundtable, International Council on Systems Engineering (INCOSE), International Foundation Process Analytical Chemistry (IFPAC), and American Indian Science and Engineering Society (AISES).

In his spare time, Tim can be found enjoying time with family, walking the family dog Buddy, playing Bluegrass guitar, or discussing the finer points of theology.

The Case Study – KIND Karz

Charlie hadn't cried in a long time—even when a distant family member passed, he struggled to shed tears. And it worried him that he was unable to adequately express his emotions. But this might do it—this was bad news on top of bad news. His second (seemingly successful) business venture was in serious trouble.

He was something of a prodigy. His first company, QUILZ, was a near instant success. His new voice recognition technology had revolutionized the finance industry—a 99.99% accurate voice recognition system that made credit cards obsolete. Just say your name in a microphone at checkout, and voila, your credit card was instantly debited with your purchase. But then—he walked away to start something new with over 5 billion US dollars in the bank.

Like the first time, they called him crazy. This time he planned to start a car company. Not just any car company, but one that balanced multiple user preferences while remaining affordable. He built it on a philosophy of kindness (to workers, the environment, and communities), breathtaking innovation, networked (technology driven), and focused on the needs of the driver. He named it KIND Karz—KIND was an acrostic for Kind, Innovative, Network, and Driver-centric. With his new fortune, he hired a core group of the best of the best talent he could find, increasing their salaries by at least 50% and promising they would be rich beyond imagination.

After only 2 years, people stopped calling him crazy after he had done the impossible. Charlie had already built an amazing selection of cars that were cutting edge technologically and ran for pennies per mile with minimal environmental impact. And they were reliable—out of the gate he offered a 100,000-mile warrantee including routine maintenance with a 15-year expiration date, transferable to other owners also. He set up a creative deal with local used car dealers (who had to first meet rigorous quality and integrity assessments) to sell and service his cars. He financed their expansions at low interest rates and standardized repair facility designs. He developed a curriculum for mechanics that ensured a highly competent workforce that was second to none. Further, local dealers were able to leverage their facilities to service other brands, and the cash was pouring in.

© The Author(s), under exclusive license to Springer Nature Switzerland AG 2022
T. D. Blackburn, *Six Sigma*, https://doi.org/10.1007/978-3-030-96213-5_1

The IPO broke all records. In just 2 years, his core team realized his promise—they were rich beyond expectations. But 10 years later, concerns began to be raised about quality. Even though service was unconditionally guaranteed free up to 100,000 miles, no one wanted the hassle. (No cars were yet outside the 15-year period, but many had exceeded 100,000 miles and owners were getting nervous.)

Charlie was starting to get bad press again. They were seeing all kinds of recalls and unexpected warrantee work—everything from airbag and brake failures, emission issues, propulsion, electronics, hardware, and others. Their cost to operate took a hit, with a 25% loss in Net Profit for the first time. And with the bad press and customer satisfaction, sales were decreasing as well. Stock prices fell a proportional 25% in just 1 year.

Something had to be done.

Also, Charlie was starting to have problems with suppliers. He had forged a partnership agreement with key suppliers, but one was giving problems—it was the airbag assembler. There, the airbags were fabricated and preinstalled in a KIND Karz interior front panel. But Charlie had noticed they were struggling to keep up with demand. As a result, KIND Karz was unable to meet demand from the dealers. Further, the assembler vendor was asking for longer notice to schedule new lots of airbags (a full 6 months prior). And to make matters worse, airbag failure and recalls were getting particularly bad press and costing a lot of money to resolve. And there were a couple of fatalities, with crushing lawsuits coming his way.

Something had to be done! He felt totally overwhelmed.

Charlie reflected—he had learned a lot in 10 years. Starting out, he knew little about the automobile industry. His only experience had really been watching his father build hot rods in the shed out back. Ironically, he had no interest in it at that time. He wasn't a fan of greasy fingernails and busted knuckles. Had he finally taken on more than he could handle? What to do? Suddenly he started hating cars. And that made him want to win even more.

So, he called an old college roommate, Albert. That was what he always did when he felt overwhelmed like this. Albert had a way of seeing the realistic but in a positive way and was always an encourager. He hadn't talked to him in a several months, but he know that he had success in a company that produced medical equipment leading their operational excellence program—perhaps he would have some ideas. In the beginning, Albert had given him some good ideas and suggested that he look at the Toyota Production System (TPS) (Liker, 2020).

Charlie, an avid reader, devoured everything he could about TPS at the time and hired a couple of their high performers. His version of TPS became the standard at KIND Karz manufacturing plants, but he had yet to roll it out at key suppliers and vendors. He also knew Albert was a bit of a nerd—always loved statistics while Charlie preferred to get lost in a good science fiction novel back in college. He knew Albert had become a Master Black Belt in Six Sigma and was in high demand. Maybe he would have some ideas.

And did he ever. Against the advice of not hiring your friend, Charlie made Albert an offer he could not refuse. But it was on the condition he would turn around the quality and vendor problems. Albert immediately began to review the company financials and readily available data as in Figs. 1.1, 1.2, 1.3 and 1.4.

So, if you were Albert what would you do?

1. What is the problem here?
2. What questions would you have?
3. Take a look at the data and financials above. What does it tell you? How might you go about analyzing the data? What other data might you need?
4. How would you determine what are the root causes?
5. How would you go about identifying corrective actions and confirming they were effective?

The following are helpful references related to the case study.

Fig. 1.1 Defect types and counts

Defect Categories	Count of Defect
Airbags	344
Brakes	487
Other Defects	20
Electronics	60
Emissions	80
Hardware	20
Propulsion	40

Month	Cars Produced	Warrantees Claims and Recalls
Jan 2018	1013	192
Feb 2018	992	195
Mar 2018	960	191
Apr 2018	996	207
May 2018	983	194
Jun 2018	955	195
Jul 2018	951	195
Aug 2018	984	204
Sep 2018	1023	206
Oct 2018	986	209
Nov 2018	967	198
Dec 2018	1017	194
Jan 2019	1005	196
Feb 2019	969	195
Mar 2019	1022	203
Apr 2019	969	208
May 2019	970	197
Jun 2019	971	207
Jul 2019	1014	209
Aug 2019	954	205
Sep 2019	1007	195
Oct 2019	1008	209
Nov 2019	964	200
Dec 2019	1000	200
Jan 2020	1016	195
Feb 2020	1017	199
Mar 2020	991	196
Apr 2020	1008	206
May 2020	1020	195
Jun 2020	968	206
Jul 2020	970	197
Aug 2020	971	207
Sep 2020	1014	209
Oct 2020	954	205
Nov 2020	1007	195
Dec 2020	1008	209

Fig. 1.2 Warrantee and recall counts

ASSETS	2019	2020
Total current assets	31151.25	21498.4
Operating lease for vehicles, net	7837.5	5524.4
Other related business	23516.25	17477.6
Property, plant, and equipment	42487.5	27582.8
Operating lease	0	3508.4
Goodwill and intangible assets, net	1312.5	974.4
Customer notes receivable	1582.5	2010
Restricted cash	1492.5	991.2
Other assets	2145	2242.8
Total assets	111525	81810
LIABILITIES AND EQUITY		
Total current liabilities	37473.75	25880.4
Debt and finance leases	35265	27406.4
Deferred revenue	3716.25	3239.6
Resale value guarantees	1233.75	590.8
Other long-term liabilities	10162.5	6932.8
Total liabilities	87851.25	64050
Redeemable noncontrolling interests in subsidiaries	2085	1596
Total stockholders' equity	18461.25	13750
Noncontrolling interests in subsidiaries	3127.5	2413.6
Total liabilities and equity	111525	81809.6

Fig. 1.3   Balance sheet (in 1000's US dollars)

REVENUES	2019	2020
Sales	18000	16000
Leasing	900	700
Services	1400	1600
Total Revenues	20300	18300
OPERATING EXPENSES		
Research	1500	800
Selling, general, administrative	14740	22,000
Total Operating Expenses	16240	22800
Net Income (Loss) from Operations	4060	-4500
Tax	1218	-1350
Net Income (Loss)	2842	-3150

Fig. 1.4   Statement of operations (in 1000's US dollars)

References

Antony, J., Snee, R., & Hoerl, R. (2017). Lean six sigma: Yesterday, today and tomorrow. *The International Journal of Quality & Reliability Management*, 1073–1093.

Federico, M., & Beaty, R. (2003). *Six sigma team pocket guide*. McGraw-Hill.

George, M. L., Rowlands, D., Price, M., & Maxey, J. (2005). *Lean six sigma pocket toolbook*. McGraw-Hill.

Liker, J. K. (2020). *The Toyota way* (2nd ed.). McGraw Hill.

Patterson, G., & Fedrico, M. (2006). *Six sigma champions pocket guide*. Rath & Strong.

An Introduction to Six Sigma

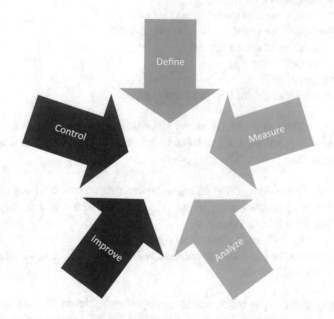

In the prior chapter, the KIND Karz case study was introduced. To summarize, KIND Karz is a new, start-up automobile company and is highly innovative. Its name is an acrostic of sorts, with KIND being made up of the terms Kind, Innovative, Networked, and Driver-centric. Charlie tried to build the company around this perspective. Specifically, he built the company on a philosophy of kindness (to workers, the environment, and communities), cars built with breathtaking innovation, deeply networked (technology driven), and with an intense focused on the needs of the driver.

And Wall Street bought it—the Initial Public Offering (IPO) broke all prior records, with the closing price greatly exceeding expectations. Good times

T. D. Blackburn, *Six Sigma*, https://doi.org/10.1007/978-3-030-96213-5_2

continued. However, after 10 years, significant concerns of quality began to emerge. The very customers Charlie had worked so hard to attract (and exceed their expectations for a car) were beginning to complain, directly related to defects and recalls. Although most so far had been covered by warrantee, it remained an inconvenience and aggravation for the car owners.

Given the recent warrantee and rework costs along with a drop in sales, stock prices are down 25% from the prior year, during an otherwise bull market. How should newly hired operational excellence wizard Albert resolve this problem? What questions should Albert ask?

- What are the issues?
- What are the trends over time?
- What data do we need to gather to sort it out?
- Which issues are the worst (so we can prioritize)?
- What is important to the customer?
- What are the root causes?
- What can we do to fix it?
- What can we do to ensure it won't occur again?

Likely you will agree the above are appropriate questions to ask. But is this really what typically happens in practice? For most of us, experience has shown otherwise. The following is what often happens.

- Jump to solutions – identifying solutions before getting to underlying root causes.
- Political-type response – to try to avoid losing face with one's manager, or customers, stockholders, or regulators.
- Rush to correct – sense of panic to stop losing customers.
- Base on opinion versus data. Speculation will abound. Often, the loudest or most senior opinion might win.

Interestingly, the types of questions a properly structured Six Sigma approach will take will more closely align with the initial list of questions. So, what is Six Sigma? Is it the same as Lean Six Sigma?

Simply stated, Six Sigma is a methodology for effective problem-solving, focusing on reducing defects and variation. It is credited as originally starting at Motorola in the 1980s by Bill Smith and Mikel Harry and derives its name from statistics (Antony et al., 2017). A process functioning at Six Sigma is expected (over time) to have no more than 3.4 defects per million opportunities (DPMO). It is a 99.99966% defect-free process.[1] See Fig. 2.1, which illustrates the extent of full plus or minus Six Sigma—this level of performance is approaching perfection.[2] This is a different

---

[1] To be precise, this is the level of defects at 4.5 sigma, accounting for an expected 1.5 sigma process shift over time.

[2] The 9.8659E-10 are the probabilities of being above or below 6 standard deviations from the mean.

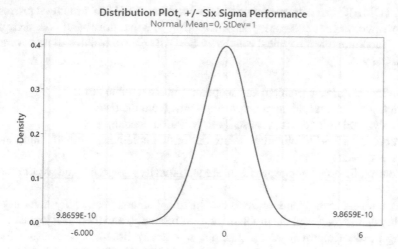

Fig. 2.1 Normal distribution at +/− Six Sigma

way of thinking versus incremental improvement and sets a high bar of performance that forces a fresh approach.

So, what about Lean Six Sigma? Often, people link Lean and Six Sigma as a common approach. While there is often overlap, there is a distinction that should be made before we continue. Also, we should understand such a distinction to ensure an effective and efficient deployment of each. As noted prior, Six Sigma focuses on identifying and reducing defects or variation. However, Lean seeks to improve flow while avoiding waste and inefficiencies. First, it focuses on reducing excess variation or defects. Otherwise, a Lean project could result in making defective product more quickly. While Lean remains another essential tool kit, this book will focus on Six Sigma (and therefore reducing variation or defects.)

Although its title is derived from statistics, Six Sigma has become more of a systematic problem-solving approach using the DMAIC (Define, Measure, Analyze, Improve, Control) phased approach. DMAIC is a systematic way of approaching difficult and novel problems. It is useful when the true root cause and solution is not known. (If the root cause and solution is truly known, DMAIC is likely note needed.) However, when used, it enforces discipline and helps overcome common biases. When not used, the following can occur:

- Jump to conclusions as to the root cause.
- Root cause by consensus.
- Lack of a data-centric approach.
- Failure to realize the risks and sometimes make problems worse.
- Problem comes back, either because a root cause was missed, or the solution was misaligned, or there were no control mechanisms to ensure it doesn't reoccur.

But DMAIC can help overcome the above. As noted, it is a structured process to help overcome natural biases when attempting to solve novel and complex problems. The following are typical actions by DMAIC phase, which will be covered in detail in later:

- Define: Define the problem and scope in understandable terms.
- Measure: Understand the problem and learn from the data.
- Analyze: Identify the root causes (verify with evidence).
- Improve: Identify aligning solutions and implement without introducing excess risks.
- Control: Ensure the improvements are statistically significant and won't reoccur.

As with most problems or decisions in life, humans necessarily naturally and quickly define (or grasp) the problem (at least mentally) and then rush to put in mitigating factors (improve). As an example, a motorcyclist must be aware and alert when another vehicle enters an intersection. A decision must be made whether to slow down or maintain the current speed. Sometimes, it is as if the person doesn't see the rider and pulls out, putting the motorcyclist at risk. Of course, the rider doesn't stop and intentionally go through DMAIC. He or she instinctively decides what to do based on experience.

Similarly, this strong fight or flee instinct can carry over to solving other complex problems that cannot be solved instantly or instinctively. To overcome this instinct, a phased approach is needed to solve novel problems and help us overcome this natural inclination. As an instinct, people often loosely (and informally) define a problem and immediately move to solution.

In applying DMAIC, it is helpful to answer fundamental questions. For Define, recall the problem and scope will be defined in understandable terms. To do that, answer the following:

- What is my problem?
- Why do I need to solve it? (e.g., business case).
- What is my baseline? Can I define my problem in the form of $Y = f(X)$?
- What is my scope? Where does it start and end?
- What does the process look like? (map it).
- What does my customer need? (could be an internal stakeholder).
- Who will help me? (team members).
- Has this problem occurred before, and what can I learn from others' experiences?
- When do I need to have it completed?
- What do I need to know about my stakeholders?

Then, avoiding the instinct to move to "Improve," first understand the problem, learning from the data. This occurs in the Measure phase, where we still resist the urge to jump to solution or even assess terminal root causes. The following questions should be considered in the Measure phase:

- Is my process stable (my Y or Critical to Quality (CTQ) factor)?
- Is my process capable (e.g., Ppk) of meeting customer specifications?
- What are potential X's?
- What data do I need to collect?
- Can I trust my measurement system? Is the variation coming from my process, or my measurement system, or both?
- Once I have some data, are there clues where to look for root causes later? Are there stratification factors?

A natural tendency at this point is to make assumptions as to root cause, but that should be reserved for the Analyze phase. In the Analyze phase, consider the following questions:

- What are potential root causes?
- What are my verified root causes?
- What evidence do I need to reject or accept the root causes?
- Do I have sufficient process analysis or data analysis (e.g., statistical significance) to proceed to the Improve phase?
- What are my conclusions from the Analyze phase, and which root causes am I carrying forward to the Improve phase?

Finally, the project arrives at the Improve phase, where the objective will be to identify and align solutions without introducing excessive risks. To do so, answer the following questions:

- What root causes carried forward that I need to address?
- What are solutions that will resolve the root causes?
- Do I have sufficient funding and a reasonable financial payback?
- If there are more solutions than time or funding can afford, how many do I need to meet my project goal or objective?
- Who will ensure they the improvements are implemented, and when must they be completed?
- What are the risks of the solutions, and how will any unacceptable risks be abated?

It is at this point another human tendency must be overcome—that is, to stop the project and claim success. However, there needs to be evidence that the desired improvements were met and won't reoccur. This is the Control phase. To help guide through this, the following questions are helpful:

- Did I meet my objective? If so, was the improvement statistically significant?
- How will we ensure the improvements are sustained?
- How will we know if we start to drift?
- How do we pass along new ways of working to employees and management?
- How do we standardize new ways of working?
- What did we learn from this project?

- Are there any carryover projects or activities that need to follow?
- To whom and how do we hand off responsibilities to own this going forward?

In coming chapters, we will review the five phases of DMAIC in a Six Sigma context in detail and include typical tools and statistical methods needed in the data-driven approach of Six Sigma methodology. The main sections describe the phases and provide examples from the perspective of a team member or stakeholder. Additional sections are provided for practitioners and will include detailed steps for how to perform the statistical analysis in Minitab.

The following are some other helpful references: (Federico & Beaty, 2003; George et al., 2005; Patterson & Fedrico, 2006).

References

Antony, J., Snee, R., & Hoerl, R. (2017). Lean six sigma: Yesterday, today, and tomorrow. *The International Journal of Quality & Reliability Management*, 1073–1093.

Federico, M., & Beaty, R. (2003). *Six sigma team pocket guide*. McGraw-Hill.

George, M. L., Rowlands, D., Price, M., & Maxey, J. (2005). *Lean six sigma pocket toolbook*. McGraw-Hill.

Patterson, G., & Fedrico, M. (2006). *Six sigma champions pocket guide*. Rath & Strong.

The Define Phase

3

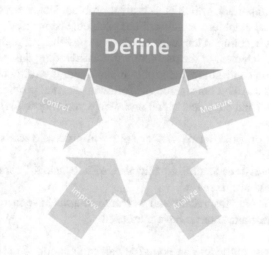

For the Define phase, recall the problem and scope should be defined in understandable terms. To do so, there are several tools at Albert's disposal. The outcomes will be included in a storyboard to document and communicate the project outcomes. These will include a Project charter, CTQ Hierarchy, SIPOC diagram, process map, stakeholder analysis, learnings from similar experiences, and background information as needed.

3.1   Project Charter

For the KIND Karz project, Albert began by crafting a Project Charter. This will address the following initial questions he wanted to answer:

© The Author(s), under exclusive license to Springer Nature Switzerland AG 2022
T. D. Blackburn, *Six Sigma*, https://doi.org/10.1007/978-3-030-96213-5_3

- What is my problem?
- Why do I need to solve it? (e.g., business case)
- What is my baseline? Can I define my problem in the form of $Y = f(X)$?
- What is my scope? Where does it start and end?

Common elements in Project Charter include a description of the project, problem statement, business impact, baseline and goals, project scope, team members and roles, and a high-level project plan or schedule.

It can be surprisingly difficult to craft a description of the project. Try to succinctly describe what the project will include and the intended objective. For KIND Karz, the project description could be, "This project will seek to identify root causes and implement corrective actions for recently observed increase in recalls and warrantee claims for KIND Karz."

Next is perhaps the most challenging thing to do in the beginning of a project but one of the most important, which is to craft a problem statement clearly and concisely. As a general guideline, a well-crafted problem statement should be one sentence and written in common language as much as possible for understandability. It should be a clear statement of the problem without including the root cause or solution. Also, early on confirm that the problem exists and is worthy of solving.

To illustrate this, Albert considered the following several problem statements for KIND Karz. Pause and consider, which one would you choose? Why?

1. Due to a lack of a risk-based vendor selection process, recalls and warrantees have increased.
2. Due to cost pressures leading to selection of substandard components, recalls and warrantees have increased.
3. Over the last 18 months, recalls and warrantee claims have increased 25%.
4. Recalls and warrantee claims have increased.

Albert chose one of the four and provided an explanation to the team as follows:

1. *Due to a lack of a risk-based vendor selection process, recalls and warrantees have increased.* Not chosen: the problem statement includes a solution (selection process) in the statement.
2. *Due to cost pressures leading to selection of substandard components, recalls and warrantees have increased.* Not chosen: the problem statement includes a perceived root cause (cost pressures).
3. *Over the last 18 months, recalls and warrantee claims have increased 25%.* Chosen: This is a clear statement of the problem (recalls and warrantee claims have increased) and other useful information included (timeframe and percent of increase).
4. *Recalls and warrantee claims have increased.* Not chosen: It is a clear statement of the problem, but without the context option 3 included.

Also, Albert considered the business impact in the charter. This helped to confirm the problem was worthy of solving and helped secure buy-in from management and other stakeholders. As a general guideline, also keep the business impact statement short (one to three sentences). In the statement, describe why it is important to resolve the problem using tangible concerns that can be identified. To help guide the crafting of the statement, consider the following categories: cost implications, compliance or regulatory concerns, safety, health or environmental impacts, ability to maintain supply, and sustain quality.

Clearly for KIND Karz, the business impact can be identified as affecting several of the categories, including cost, regulatory, safety, and quality. Specifically, Albert crafted the business impact statement as, "Increased cost and declining demand could continue if not resolved, and fully addressing quality and safety concerns are part of KIND Karz values."

The next thing to consider in the project charter are the baseline and goals. The baseline addresses where are you today. It is helpful to think of your problem as $Y = f(X)$, with the Y being what is important to resolve. This is also referred to as the CTQ, or Critical to Quality attributed. It must be quantifiable and measurable. In the case of KIND Karz, the Y is the number of warrantee and recall claims. It is usually recommended to have one (no more than two) Y's for a project to make it manageable. For KIND Karz, Albert calculated the baseline as follows: current warrantee or recall claims per 1000 automobiles sold after 12 months—200.

Next determine the goal of the project. The goal should be challenging but realistic. The project owner (the person with overall responsibility of the process) should be consulted and included in setting the project goal. Remember this will be your measure of success for your project. While it would be ideal to have no warrantee or recall claims, Albert realized that was not possible. After consulting with Charlie and other management, it was decided that the goal would be no more than 100 warrantee or recall claims per 1000 automobiles sold. However, it is also helpful to put a stretch goal. For KIND Karz, the team decided on 50 per 1000 automobiles sold.

Also important is defining the project scope in the charter. This identifies where the project will start and end. A challenge here is to avoid an overly broad scope that could lead to the project never being completed or a too narrow scope such that the goal would not be achieved. Later, we will see the scope also should align with the SIPOC. For the project scope, Albert and the team decided on the following: "Includes suppliers of components associated with recalls and warrantees to assembly in KIND Karz plant and to Dealer/service locations."

The next element to be considered in the project charter is the project plan or schedule. This answers the question, "When do I need to have it completed?" Often a traditional bar-chart is used, with simple bars. Show by DMAIC phase. Be aggressive but realistic. Typical times vary, but 6–12 months are common for a project, although some might be shorter (or longer) depending on the urgency or scope. See Fig. 3.1 for a KIND Karz example:

The final element in the charter is the identification of team members. This answers the "Who will help me?" question. While the prior elements are essential,

Task Name	Duration	Start	Finish	F	Qtr 1, 2020			Qtr 2, 2020			Qtr 3, 2020		
					Jan	Feb	Mar	Apr	May	Jun	Jul	Aug	S
Define	1 mon	Mon 1/27/20	Fri 2/21/20										
Measure	2 mons	Mon 2/24/20	Fri 4/17/20										
Analyze	1 mon	Mon 4/20/20	Fri 5/15/20										
Improve	2 mons	Mon 5/18/20	Fri 7/10/20										
Control	2 mons	Mon 7/13/20	Fri 9/4/20										

Fig. 3.1 Project plan

Table 3.1 Team members

Role	Name	Function
Process owner	Rebekah N.	Mfg operations
Sponsor	C. S. Jones	Quality systems
Project lead	A. Abernathy	OpEx
Team member	L. W. Smythe	Mfg operations
Team member	C. M. Martin	Supplier quality
Team member	A. W. Gonzalez	Repair operations

the proper selection of team members is crucial to the successful outcome of a project. First, keep the team small—typically four to six team members are recommended plus the lead role. More people than that can lead to inefficiency and difficulty in making decisions.

Ensure major functional areas have a representative. Also, when possible, choose people who want to be on the team—that will make the facilitation much easier. Passionate team members need little motivation to remain engaged and participate. Avoid political appointees and ensure someone is on the team that knows the process well. Finally, ensure expectations for the team members is understood and that their management is willing to allow time for their participation. The following are the typical team roles and responsibilities: (Patterson & Fedrico, 2006).

- Project owner: Overall identified owner of the process. Ensure buy-in and resources are available.
- Project sponsor: Monitor progress. Redirect as needed. Ensure roadblocks are removed.
- Project lead: Overall Six Sigma lead, data analysis, facilitate meetings, drive to completion.
- Core team (four to six people).
  - Attend every meeting as much as possible.
  - Coordinate with others in their functional area.
  - Complete assignments on time.
- Other key support: Identify others who need to provide key information or support to ensure project success.

For KIND Karz, Albert identified the team members as in Table 3.1.

3.2   Voice of the Customer (VOC) and Critical to Quality (CTQ) Hierarchy

After the charter, Albert began to consider the Voice of the Customer (VOC) and what was the critical parameter (CTQ or the Y) of the project. The purpose of this next step was to go from the Voice of the Customer to measurable criteria and further identify and clarify the within-scope focus (George et al., 2005, pp. 55–63). This will answer the question, "What does my customer need?" This exercise begins with the Voice of the Customer (VOC)—that is, what is important to the customer. This should be a clear statement as to what the customer wants. Then decompose the VOC further to the drivers needed to achieve the VOC. Continue to decompose the problem until measurable criterion is established and the CTQ is identified. This will be measurable in terms of your Y, and usually there will be other Y's as well. However, indicate what is in scope for your project. For the KIND Karz project, see Fig. 3.2.

Note the VOC, CTQ hierarchy is like an organization chart but rotated 90°. See Fig. 3.2 for an example. The VOC is the general criteria the customer is looking for. To arrive at this, focus groups could be formed, surveys issued, or organizations consulted that represent the customers (e.g., trade or consumer interest organizations). In the case of KIND Karz, it is somewhat obvious—the customer desires to have a reliable car. But how does one measure reliability? First consider drivers or the elements the next level down. For reliability, the customers need performance and safety. But as before, how are these measured? Decompose further to the next level, where measurable factors can be identified. For this example, these include warrantee claims related to power train, efficiency, safety equipment and prevention, and braking and acceleration. In this example, the CTQs be even further

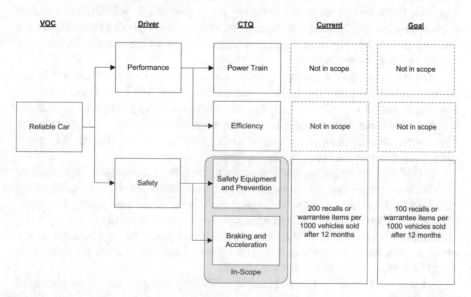

Fig. 3.2 VOC, CTQ hierarchy

decomposed to arrive at more granular and measurable attributes. There also might be others depending on the situation. But note the ones in scope are measured at least in terms of warrantee and recall claims (the focus of this project), aligning with the baseline identified in the project charter. Also, the goal is included.

Next in the Define phase, consider mapping to address the question, "What does the process look like?" The most fundamental mapping that should be included in the Define phase is the SIPOC.

3.3   SIPOC and Mapping

At first glance or use, it might be difficult to see the value of the SIPOC chart. However, it is needed and is especially helpful for Six Sigma projects where the lead might not be familiar with the process. The term "SIPOC" is an acronym for Supplier, Inputs, Process, Outputs, and Customer (George et al., 2005, p. 38). Suppliers represent the providers of any inputs to the process. Inputs are converted to Outputs. This also can be helpful identifying potential X's (predictors) or stratifications factors. "Process" is a simple map of the process flow, consisting of three to seven steps only. Important note: This should match the start and stop identified in the scope section of the Project Charter. Next are Outputs, or deliverables of the process. This can be beyond just the product itself and can include such things as testing results, service, and many other important deliverables. Last is Customer, which includes the groups or people who receive the Outputs. This can be especially helpful for stakeholder analysis that will be discussed further.

For the KIND Karz example, after creating the Process map, Albert continued in this order: he identified Outputs, then Customers, then Inputs, then Suppliers. See Fig. 3.3. Note the customers are certainly the purchasers of KIND Karz but also include distributors, dealers, and other stakeholders. Also note that there are several vendors in the Suppliers category, especially focusing on the areas of concern for recalls and warrantee work.

Also consider other mapping. Sometimes other mappings are included in the Measure phase but might also be helpful in the Define phase if it adds to understanding. Traditional flow charts are helpful, as well as swim lane diagrams (which are also flowcharts but with assigned areas of responsibility) (George et al., 2005). See Fig. 3.4 for an example of a swim lane map for KIND Karz, related to the team's early concerns over the brake and airbag recalls.

As guidance for mapping, it is important to walk the process to understand how the process operates (versus the theoretical or prescribed method). The map should accurately reflect the as-is state of the process. Also look for complexity and hand-offs, which frequently can be areas where problems or error occurs.

Mapping is not limited to the Define or Measure phase but can be used in any phase to enhance understanding. Maps can also be modified to show the to-be state for implementation, as well for training aids.

Next on the list for the Define phase is to identify and gain an understanding of stakeholders.

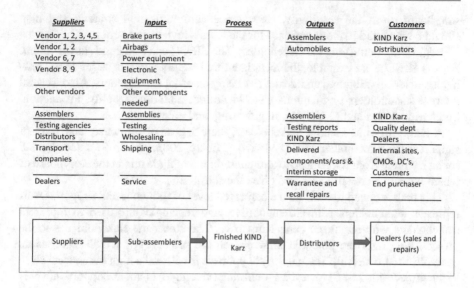

Fig. 3.3 KIND Karz SIPOC

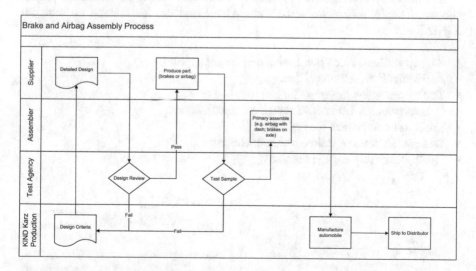

Fig. 3.4 Brake and airbag assembly process swim lane diagram

3.4   Stakeholder Analysis

Stakeholder analysis is crucial to identify key influencers of the project and who can have an impact on the project success. This addresses the question, "What do I need to know about my stakeholders?" A stakeholder analysis will include an assessment of who or which groups will have an impact on the project and the degree of support for the project. Tools used or notes taken should not be included in the project storyboard given the sensitivity. However, a communication strategy and plan should be included in the storyboard for the team's review and action.

Begin with identifying the stakeholders. The SIPOC can be a good starting point. For the KIND Karz example, the team identified supply vendors, assemblers, testing agencies, distributors, and dealers as being external stakeholders. Also, several internal stakeholders or functions were identified, including Quality, Production, Purchasing, Sales and Marketing, Engineering, and others.

A helpful tool to identify stakeholders, their degree of influence, and level of support is the Stakeholder Map (Federico & Beaty, 2003, pp. 87–91). See Fig. 3.5 for an example from KIND Karz. Reminder: do not include this in the storyboard or where it could be readily accessed given the sensitivity.

This map is coded to indicate the perceived level of influence and support. Ovals represent stakeholders who are supportive, and rectangles are used to represent stakeholders who might *not* be as supportive. The closer the stakeholder is to the project, the grater the potential influence on the project. For KIND Karz, as an example, the Leadership Team is very supportive (oval) and a high influencer (close to the project title box). However, Purchasing (rectangle) is not as supportive, yet is also an influencer. Both will be considered in a stakeholder strategy.

The next step after drawing the map is to consider strategies needed for project success. The following are some ideas Albert and the team came up with for KIND Karz:

- Supply vendors: Need purchasing help to get support.
- Assemblers: No action needed.
- Testing agencies: Leverage for test trends.
- Car distributors: Updates of status, communications.
- Transport companies: No actions.
- Dealers: Updates of status, communications.
- Quality: Actively engage regularly.

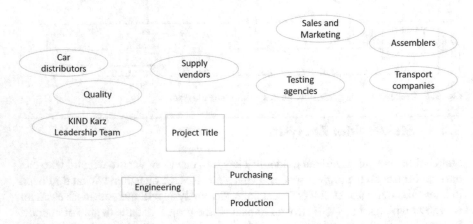

Fig. 3.5 KIND Karz stakeholder map

Table 3.2 Communication plan

Name or group	Frequency	Method
KIND Karz leadership team	Monthly	In LT staff meetings, brief update by sponsor
Key suppliers	Monthly, or as new learnings emerge	Verbally by POC
Distributors and dealers	Monthly	In writing
KIND Karz quality, sales/marketing	Monthly	Email updates
KIND Karz production, engineering, purchasing	Monthly	Direct meetings

- Production: Direct team involvement, frequent updates.
- Purchasing: Frequent updates, show impact financially.
- Sales and marketing: Need their help with communications.
- Engineering: Active team engagement—add them to the core team.
- KIND Karz leadership team: Need in-person updates.

Next, complete a communications strategy based on the learnings from the Stakeholder Map. Include the strategy in the storyboard, although be careful in how the strategies are written given potential sensitivities. Be specific as to the approach and avoid terms like "as needed." See Table 3.2 for a Communications Plan identified by Albert and the team.

The Stakeholder Analysis should be revisited and updated early in each phase. Things change over time, and an individual or function that might be supportive early might not be later. For example, when transitioning to the Analyze phase, those who were supportive prior could become anxious if a process they are responsible for is found to be at fault.

Lastly, look for and learn from other similar situations that might have occurred in the company (or elsewhere) before moving to the next (Measure) phase.

3.5   Learnings from Similar Incidences

There is one remaining question that needs to be answered in the Define phase: "Has this problem occurred before, and what can I learn from others' experiences?" Before wrapping up the Define phase, look for similar issues within your company and broader industry. What were other key findings? Can you gain any insights that could be helpful? As a watch-out, don't assume the same root cause as occurred in a similar project. Instead, use learnings as clues as to where to look. Also learn how the prior project evaluated the situation, key findings, and ideas that might be applicable in the current project. In some rare cases, the problem might have already been solved if the situations were sufficiently similar. For the KIND Karz case, look for other recall and warrantee claims. Are there any learnings that can help with this project? See Table 3.3 for some for the results of Albert's evaluation of similar prior projects and issues.

Table 3.3 Learnings from other events

Issue	Affected group or company	Learning
Recall 12 months ago for airbags	Vendor 2 (same as KIND Karz uses)	Indicated the cause was a faulty firing mechanism
KIND Karz type 4 vehicle recall due to lawsuit	KIND Karz, class action lawsuit for injuries	Wheel bearing failure
Brake assembly recall	Competitor car company	Brake pad material deterioration
Steering recall, rack, and pinion failure	KIND Karz, with type 4 car the most impacted	Unsure of cause of failure, but higher tensile gear replaced

3.6   Other Background Information and Contents

In most cases, the content described up to this point are sufficient for the Define phase. However, there might be other background information needed. These could include a timeline of events, differentiation of product, or others as needed. Photos can also be helpful. Below are some other background information and photos (Figs. 3.6 and 3.7) that could be helpful for the KIND Karz Define storyboard:

- Recalls and warrantee claims have included variety of reasons or systems, including emissions, electronics, propulsion, hardware, brakes, airbags, and other defects.
- Up until 12 months ago, KIND Karz was averaging 160 recalls or warrantee claims per 1000 cars sold in the first 12 months of ownership. That has now increased to 200.
- The most severe consequences have been brakes, resulting in a class action lawsuit.
- There have been no reported incidences with airbags, other than a fault light coming on, but one dealer noted there was torn bag fabric after an accident.
- All other recalls have been initiated by the supplier or assembler.

Figures 3.6 and 3.7 are photos to illustrate some of the issues observed, forwarded from the dealer repair shops.

In conclusion, the Define phase should identify the problem, scope, and baseline and provide another context needed for project success. Next, we will continue to the Measure phase.

The following are some other helpful references related to the Define phase: (Federico & Beaty, 2003; George et al., 2005; Patterson & Fedrico, 2006).

Fig. 3.6 KIND Karz
brake caliper failure

Fig. 3.7 KIND Karz
airbag seal failure

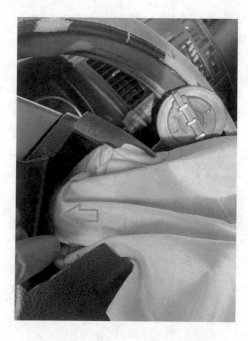

References

Federico, M., & Beaty, R. (2003). *Six sigma team pocket guide*. McGraw-Hill.

George, M. L., Rowlands, D., Price, M., & Maxey, J. (2005). *Lean six sigma pocket toolbook*. McGraw-Hill.

Patterson, G., & Fedrico, M. (2006). *Six sigma champions pocket guide*. Rath & Strong.

The Measure Phase (with Minitab Tools)

4

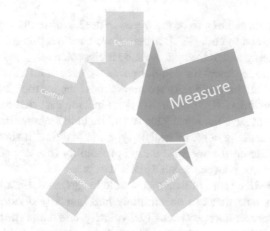

4.1   The Measure Phase: An Overview

As we transition from the Define phase to the Measure phase, we need to avoid the natural human instinct to jump ahead to the Improve phase for solutions. In the Measure phase, we will pause to better understand the problem and get clues from the data. By the time we complete the Measure phase, we would like to have confidence in our data, know whether the process is stable and capable, and have clues as to where the root cause might be before we transition to the Analyze phase. A caution also in the Measure phase is to avoid the natural tendency to look for the terminal the root causes.

Electronic Supplementary Material: The online version contains supplementary material available at [https://doi.org/10.1007/978-3-030-96213-5_4].

Typical contents of the Measure phase storyboard are as follows, which we will develop for KIND Karz in this section:

- Additional process maps if needed.
- Potential predictors (X's) – factor priority matrix.
- Data collection plan.
- Measurement system analysis (MSA).
- Process stability.
- Process capability.
- Stratification.
- Summary of the measure phase.

But let's first start with considering what data we need to collect.

4.1.1   Data Collection

Recall the questions needing to be answered in the Measure phase, one of which is "What data do I need to collect?" But before we consider this for KIND Karz, the most important thing is that the data must be representative.

First, it needs to represent the population. In the case of KIND Karz, the population would reflect all the vehicles in the market in scope for the study. We also need to account for the range of variation. This can include cars that have no warrantee claims to those that have the most. This would include the timeframe of interest. It also should represent the range of vehicle types. In addition, it should be of a sufficient sample size to ensure we have a high probability (usually 80% or higher) of detecting a difference between recall rates if it exists.

Next consider what types of data that should be gathered. First, there is continuous data. This is data that can be infinitely and logically divided, such as time, money, weight, temperature, and density. Typically, less data is needed for continuous data versus other forms, and analysis methods are somewhat easier to understand and learn.

Also, discrete data is common. This includes count data, where it doesn't make sense to divide it (e.g., one cannot have half a defect). Another type of discrete data is binomial, where only two outcomes are possible such as pass or fail, 1 or 0, yes or now, defect or non-defect.

There are other types of data, such as nominal data (e.g., a label of some type), with no logical order. This includes things like colors (red, yellow green), machine 1, 2, 3, etc. Attribute data is a type of nominal data and is commonly used to identify and quantify defects. Ordinal data, such as occurs in a Likert scale, is ordered data (e.g., 1, 2, 3, 4, 5) from high to low.

But with any data type, be aware of cautions. Usually, we are provided historical data from one or more sources. The data might not be in a form that can be used and could require modification (without changing the original data). And it might not be sufficiently representative of the population. And the challenge with data provided to us is we might not know when or how the data was collected and might not understand it. Then there is the question of whether the data was collected properly, and if it is of sufficient quantity and variation to be useful. So, these need to be considered when acquiring or gathering data.

But what data should we choose for our project? A helpful tool in Six Sigma is a factor prioritization matrix. This is a helpful tool to brainstorm initially what is expected to be significant factors related to our project. Start with identifying the Y's, and the predictors (X's) or stratification factors (also a kind of X). A spreadsheet such as Excel can be helpful, scoring Impact and Severity of the factor. Then factors with the highest scores should receive priority and transfer to the data collection plan.

See Table 4.1 for an example from KIND Karz. Note the Likelihood of Impact and Severity of Impact allow for a scoring of 1–10 and the importance score is a product of the two. These identify the initial thinking as to which X's to gather and answer another Measure phase question, "What are potential X's?" Also note there are two Ys in this case—warrantee and recall claims per 1000 cars sold, which will be merged later (for this example) in the data collection plan.

Table 4.1 Factor prioritization matrix

Factor description	Response or predictor?	Likelihood of impact on Y (1–10)	Severity of impact on Y (1–10)	Importance score	Include in data collection plan?
Warrantee claims first 12 months per 1000 cars sold	Response (Y)				
Recall claims first 12 months per 1000 cars sold	Response (Y)				
Claims per vehicle type	Stratification factor (X)	7	7	49	Yes
Claims per factory location	Stratification factor (X)	7	7	49	Yes
Claims per component type	Stratification factor (X)	8	8	64	Yes
Claims per component vendor	Stratification factor (X)	8	8	64	Yes
Airbag tensile strength	Predictor (continuous) (X)	9	9	81	Yes
Brake caliper torsion	Predictor (continuous) (X)	7	9	63	Yes
Fluid pump pressure	Predictor (continuous) (X)	1	8	8	No
Airbag defects	Predictor (discrete) (X)	9	9	81	Yes

The data collection plan comes next and should occur early in the Measure phase to describe what data needs to be collected. It also encourages a kind of discipline for the team as well and reinforces the need to be data-driven.

As noted prior, this represents the early thinking as to what data needs to be collected. While there will be learnings throughout the project that could lead to a need for additional data, be as comprehensive as possible early, as the providers of the data can become weary with repeat requests.

The data to be collected should be based on the factors carried forward from the priority matrix. Table 4.2 is an example for KIND Karz. Note that each factor is indicated as an Y or an X, and the factor is described. Then the type of data is indicated, as well as the source of the data. Also important is the timeframe of the data to be collected. The data must be relevant and representative. For example, does the data represent a timeframe where the process was relatively unchanged? If mixed with a major change, the data could be misleading. Finally, indicate how the measurement system will be evaluated (this is key to determining whether we can trust the data)—this will be covered later in this section.

Table 4.2 Data collection plan

Factor description	Response or predictor?	Data type	Source	Timeframe	MSA method
Warrantee claims first 12 months per 1000 cars sold	Response (Y)	Proportional	Quality defect database	36 months	Operational definitions
Recall claims first 12 months per 1000 cars sold	Response (Y)	Proportional	Quality defect database	36 months	Operational definitions
Claims per vehicle type	Stratification factor (X)	Discrete (counts)	Quality defect database	36 months	Operational definitions
Claims per factory location	Stratification factor (X)	Discrete (counts)	Quality defect database	36 months	Operational definitions
Claims per component type	Stratification factor (X)	Discrete (counts)	Quality defect database	36 months	Operational definitions
Claims per component vendor	Stratification factor (X)	Discrete (counts)	Quality defect database	36 months	Operational definitions
Airbag tensile strength	Predictor (X)	Continuous	Lab test	Recent samples	Operational definitions
Brake caliper torsion	Predictor (X)	Continuous	Lab test	Time of gage	Gage R&R
Airbag defects	Predictor (X)	Attribute	Visual inspection	Time of AAA	Attribute agreement analysis

As to the sources of data, it is common to have access to historical data trended by the company. Often software systems include a historian system, or at least include printed records over time. But data can also be acquired through observation (such as time studies). However, the observation method should be carefully planned to ensure the operators don't perform differently due to being watched. In some cases, it is necessary to generate data, such as by a Design of Experiment (DOE) which is covered later in this book. In some cases, data might need to be simulated (i.e., Monte Carlo), but the underlying characteristics of the data distribution must be known. Also, surveys or maturity assessments can be used but are less common in Six Sigma.

Considered is how much data should be acquired. Minitab provides Power and Sample Size tools for several statistical methods, at Stat/Power and Sample Size—see Fig. 4.1. This will be addressed in specific sections of the book for the statistical method being considered. Sample size should be considered early in the data collection planning process and requires an awareness of the statistical test or method

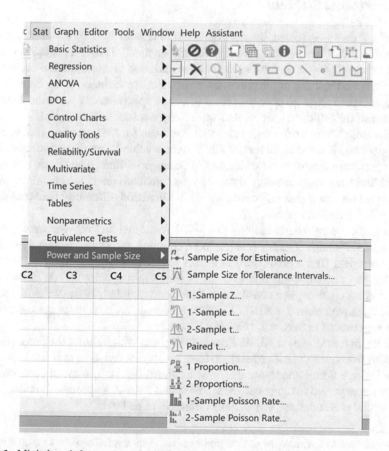

Fig. 4.1 Minitab path for power and sample size

likely to be used. It also requires some level of data understanding, such as the standard deviation, and how much difference needs to be detected.

Industry practice standards or rules of thumb are available and will be discussed where the statistical method is being described in the text. Many of these are available through Minitab's help feature.

When gathering data, there are several options to consider. The first is to use all the data available. But in some cases, the dataset might be too large to be practically manageable. If this occurs, the second option is to randomly sample from the data available. If the data is being gathered from a high-volume production process, another option is to do stratified sampling. This could be sampling a fixed quantity on a cycle, or time between sampling. Another could be systemic sampling for low-volume processes, such as automobile manufacturing. Again, ensure the data is representative of the larger dataset or population of interest.

Once we have the data gathered, the next step is to determine if the Y(s) is stable.

4.1.2   Process Stability

Once the Y data is gathered, the next question to answer is as follows: "Is my process stable (my Y or Critical to Quality (CTQ) factor)?" This is done using control charts by identifying if we have special cause variation versus common cause.

Variation is the voice of the process. Common cause variation is the normal variability expected from a stable process. However, special cause variation is something out of the ordinary, a signal that something unusual has happened.

A control chart is a time series chart with a mean (or average) line added, along with upper and lower control limits. The Y axis is value being measured, and the X axis represents chronological order or the passage of time. The upper and lower control limits are approximately three sigma or standard deviations from the mean. Note, however, the sigma for control limits is calculated differently than traditional descriptive statistical means.

Common control charts include the I-MR (or individual moving range) chart, Xbar-R, and P-charts. The I-MR is the most common and works especially well for continuous data. The Xbar-R chart is useful subgroup sampling, and the P-chart is used for proportional data. The U-chart is sometimes required when defect attributes can be counted, but non-defects cannot be counted. Next, we will look at several examples from the KIND Karz case. Details on how to create the control charts are covered in Sect. 4.2, "Control Charts."

For the first example for KIND Karz, see Fig. 4.2, which shows the proportion of cars with warrantee claims or recalls. Here, no special cause signals are indicated, and points are within the upper and lower control limits. The mean proportion is 20% but is expected to range between 16.4% and 24%[1]. The process is stable, even if we would prefer that the proportions defective to be lower.

---

[1] For the advanced users, note the control limits appear to be atypically wide. In such a case, a P-chart diagnostic is advised. This is beyond the scope of this book, however.

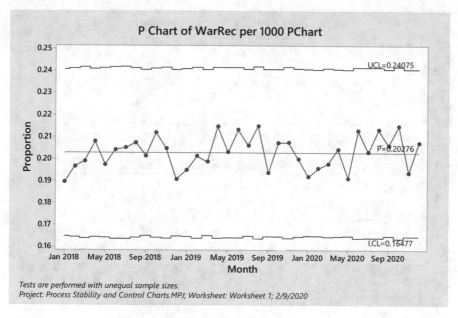

Fig. 4.2   P-chart: proportions of cars with warrantee claims or recalls

But Albert wants to study the data further. Based on the data collection exercise and feedback from operators, he wants to review brake caliper torsion rod resistance values. (Note: This is a fictional caliper component for purposes of illustration.) Since torsion resistance is a continuous factor, he chooses the I-MR chart (see Fig. 4.3). Like before, the process is stable (there are no special cause signals in the moving range or the individuals chart). The torsion readings are expected to average 77.8 and range between 74.6 and 81. The upper or individuals chart plots the actual torsion resistance values from testing, and the lower or moving range chart plots the moving range. There is nothing unusual over time for the torsion rod values, and the process is stable.

But Albert is concerned whether the data is fully representative and asks the vendor/supplier to do systematic sampling over a 2-week period. He asked them to pull three calipers every 30 min and test the torsion resistance. This requires the use of an Xbar-R chart (see Fig. 4.4). The top chart is the averages of each pull, and the bottom represents the ranges within the subgroup. There is a special cause event occurring in subgroup 10—as shown in the upper chart, it is below the lower control limit. Also, it is above the upper control limit on the bottom chart, indicating there is a greater variation in the three samples at pull 10 versus the others. This will need to be further evaluated, and the process is not considered stable.

Like the brakes, Albert suspects airbags are also a problem, so he gathered data for airbag defects. This required the use of a different kind of chart—a U-chart (see Fig. 4.5). This is useful when the lack of a defect can't be counted, but the opportunities for a defect vary (e.g., different quantities produced). For example, the airbags are checked for how many defects each have but cannot be checked for how

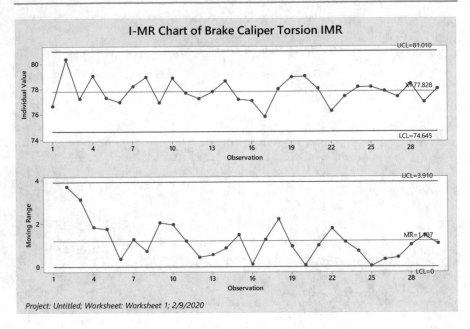

Fig. 4.3 I-MR chart: brake caliper torsion

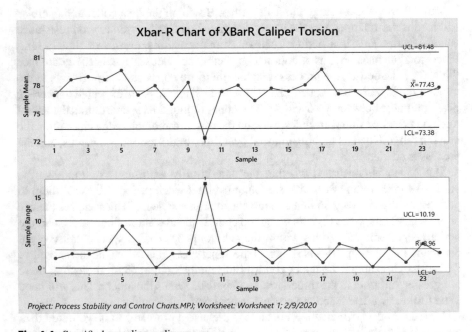

Fig. 4.4 Stratified sampling, calipers

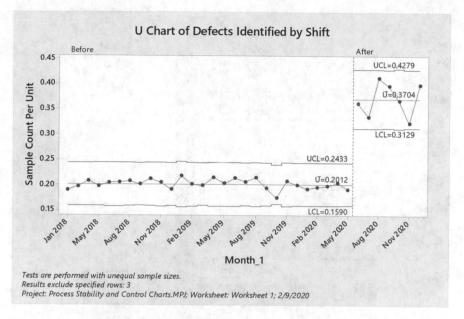

Fig. 4.5 U-chart: airbag defects

many defects they don't have. Different airbag quantities are produced daily, so the opportunities for defects will vary. For example, on days more airbags are produced, the likelihood that there will be more overall defects increases. Albert immediately noticed a shift in the data sometime mid-2020. After further discussions, he discovered this was the same time a new assembler machine was installed at the supplier. This will require further evaluation in the Analyze phase.

Some process instability was discovered for KIND Karz. This is the level of detail that is normally provided for a project participant or stakeholder. But for the person performing the analysis, see Sect. 4.2 "Control Charts" for more details and instructions on how to create in Minitab and interpret control charts as well as process stability.

But a question lingers—"Is my process capable of meeting customer specifications?" Even when stable, will the normal range (between upper and lower control limits) satisfy my customer?

4.1.3 Process Capability

In the prior section, we reviewed process stability, which is the voice of the process. But are our processes capable of meeting customer needs? That is different question. That is, will the process operate within the upper and lower specification limits (USL, LSL) which are set by the customer and not the process? In the Measure phase, it is crucial to know whether our process is capable of meeting customer needs.

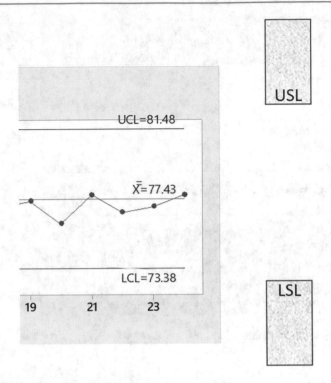

Fig. 4.6 Control limits versus specification limits

To understand the relationship of control limits and specification limits, see Fig. 4.6. A common index used for long-term process performance is Ppk. See Sect. 4.3 Capability Analysis for more details and instructions on how to calculate Ppk. But in the simplest terms, a Ppk of 1 would mean the USL = UCL and LSL = LCL. Remember that the control limits are three standard deviations from the mean, accounting for most of the expected variation (99.7%) in a stable process. That is, the process is just capable when Ppk = 1. However, in practice most processes should function at a Ppk of 1.33 or greater, which would indicate the UCL/LCL fit well within the USL/LSL. Then for a stable process, the process would be controlled well within the specification limits.

There are two primary methods this book will demonstrate to calculate process capability. The first is for proportional data (e.g., defect counts) and the other for continuous data. This is normally considered for the *big Y* (for KIND Karz total warrantee/recalls), as well as the *little Y's* (e.g., other factors of concern such as the brake caliper torsion rods and airbags).

For the *big Y*, recall the percent of cars requiring warrantee or recalls is just over 20%. The resulting Ppk is 0.28 which is less than 1 and 1.33, indicating the process is not capable of meeting expectations (see Table 4.3).

For the *little Y's*, we can start with the airbag example we used prior. There, we observed 370 defects with 1000 airbags produced, and there were 5 possible defects per airbag. See Table 4.4 for the Ppk calculation, which reveals the process is not capable with a Ppk = 0.48.

Table 4.3 Ppk for warrantees and recalls

Category	Entry or results
Proportions defective	20.276%
Process sigma	2.33
Ppk	0.28

Table 4.4 Ppk, airbags

Category	Entry or results
Number of units	1000
Defects	370
Defect opportunity per unit	5
Defects per opportunity (DPO)	0.074
Z (DPO standard deviations from the mean)	1.447
Ppk (Z/3)	0.482

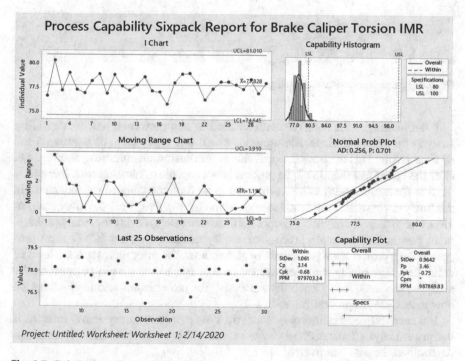

Fig. 4.7 Ppk, brake caliper torsion rods

Albert also calculated Ppk for the brake caliper torsion rods. As this is continuous data, Minitab was used to generate the analysis (Stat/Quality Tools/Capability Six-pack) (see Fig. 4.7). While the assumptions were met for data normality and stability, the Ppk is −0.75 which is less than 1, so the process is not considered capable of meeting customer needs for brake caliper torsion resistance.

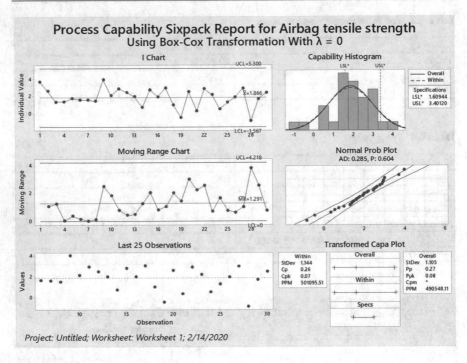

Fig. 4.8 Ppk, airbag fabric tensile strength

After looking at the data, the team decided to also perform a process capability analysis for airbag fabric tensile strength, which was expected to be between 5 and 30 k-oz. (Note: This these are fictional values for illustration purposes.) See Fig. 4.8. Note the Ppk is less than 1 and the process is not capable of meeting customer specification for tensile strength. This will need to be further evaluated for root cause in the Analyze phase. It is important to understand the low Ppk is not the root cause but gives clues where we might dig deeper for the root cause(s) during the Analyze phase.

In conclusion, there are process capability concerns for key KIND Karz processes. As a project team participant or stakeholder, the prior represents the level of detail that is normally provided and explained. But as the person performing the analysis, see Sect. 4.3 "Capability Analysis" for more details and instructions on how to analyze for process capability.

But another question lingers—"Can I trust my measurement system?" Imagine if the prior analysis indicated the processes are stable and capable, but later it was determined the data was in error.

4.1.4   Measurement System Analysis (MSA)

The next question we need to answer is, "Can I trust my measurement system? Is the variation coming from my process, or my measurement system, or both?" This

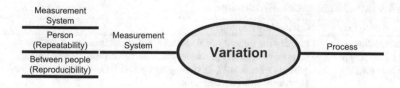

Fig. 4.9   Sources of variation

is an important question to answer, as variation in the process could be coming from the process or the measurement system. Variation within the measurement system can be from the person performing the test (repeatability), or between people (reproducibility), or the measurement device or system. Or the variation can come from a combination of all three. See Fig. 4.9 which illustrates the sources of variation. Remember a typical objective of a Six Sigma project is to reduce variation.

Measurement system effectiveness becomes especially a concern when data is gathered (or assigned) by a human. This can be when a person takes a measurement or makes attribute decisions for defects. Given the additional components of uncertainty, we need to determine if the measurement process is robust such that it sufficiently and accurately measures variation in the process. For example, if the same person measures or assigns attributes repeatedly, do they get the same results? When different people measure or ascribe attributes to the same object, are the outcomes the same? These are important questions to answer, but first we need to have confidence what we interpret is correct.

This confidence is provided in part through Operational Definitions. These are clear, precise definitions for factors, or descriptions as to how measurements were taken. It ensures consistent understanding of the factor or how it is calculated and enables correct interpretation of the data. When writing Operational Definitions, be sure to review with a subject matter expert (SME) who is familiar with how the data is gathered or created the field if in a database. For KIND Karz, Albert created the following Operational Definitions, which he confirmed with people who were responsible for gathering the data (see Table 4.5).

In cases where humans are measuring items or assigning attributes, additional analysis is usually needed. A common method for continuous data is Gage R&R, with the R&R referring to repeatability and reproducibility. Another method is the Attribute Agreement Analysis, which is used when humans are ascribing attributes (e.g., identifying defects). Examples for both will be provided in the following. The information in this section will be provided similar as occurs in a project as related to a team member or stakeholder's understanding and scope of interest. However, if you are interested in learning how to do the analysis, see Sect. 4.4 "Measurement System Analysis."

First, let's consider Gage R&R. As noted, prior, this is used for continuous data and can help provide confidence that the measurement system is reliable. It will identify issues with repeatability (same operator repeating a measurement) and reproducibility (a different operator measuring the same thing). To illustrate this method, we will determine if the measurement system for the brake caliper torsion

Table 4.5 KIND Karz operational definitions

Factor description	Operational definition
Warrantee claims after 12 months per 1000 cars sold	Identify the quantity of warrantee (not recall) claims in the first 12 months after the purchase of a new vehicle and divided by the quantity of cars sold. Divide again by 1000. Date range is the last 36 months from current date
Recall claims first 12 months per 1000 cars sold	Identify the quantity of recalls (not warrantee claims) in the first 12 months after the purchase of a new vehicle and divided by the quantity of cars sold. Divide again by 1000. Date range is the last 36 months from current date
Claims per vehicle type	For the data included above total number of claims (warrantee and recalls) by vehicle type
Claims per factory location	For the data included above total number of claims (warrantee and recalls) by KIND Karz assembly factory location
Claims per component type	For the data included above number of claims by major component categories as included in the quality Data Base tracker
Claims per component vendor	For the data included above number of claims by component vendor as included in the quality Data Base tracker
Airbag tensile strength	Determine the maximum tensile strength (Mpa) of the airbag material at the connection point. Based on 50 random selected airbags that failed, compared to 50 random selected airbags that did not fail during the period
Brake caliper torsion	Determine the maximum torsion strength (N-m) of the brake caliper at the smallest diameter. Based on 50 random selected calipers that failed, compared to 50 random selected calipers that did not fail during the period
Airbag defects	For airbags that failed, identify the attributes that do not meet standards, such as web thickness, discoloration, pinholes, missing adhesive, and overlap

is sufficient. This is an important measurement as it is a critical parameter to ensure the calipers won't fracture when braking. As such, it is the last line of defense from the vendor before KIND Karz installs the brakes on the final automobile.

After completing the study and analysis, Albert shares the following graph (see Table 4.6) with the team and explains most variation is in the parts (calipers) versus the measurement system (see the upper left-hand graph). That is good, as it suggests the measurement system is detecting a difference, versus wide variation in the testing method. He explains that the reproducibility, repeatability, and overall gage are less than 30% (the first three bars) which suggests the measurement system is reliable. Also, there is no apparent interaction between people and parts—he explains that the lines on the lower right-hand graph are nearly parallel which indicate no interactions. There are other things to consider if you are the analyst, which are covered in more detail in Sect. 4.4.

"But I do have a concern," Albert said. He then shows some Minitab output information, as shown in Table 4.7. While most variation is in the parts or brake calipers, and the percent study variation is less than 30%, the measurement system is only capable of distinguishing four categories. Good practice recommends at least five categories. We will need to further evaluate why the measurement system does not have sufficient resolution later.

Table 4.6 Gage R&R report graphs

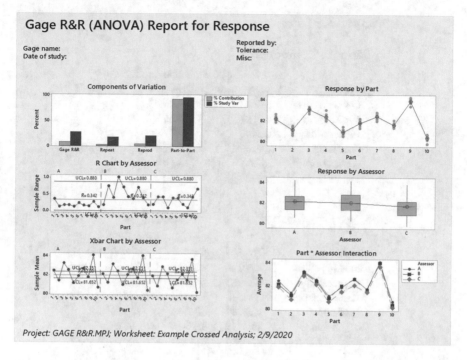

Gage R&R (ANOVA) Report for Response

Project: GAGE R&R.MPJ; Worksheet: Example Crossed Analysis; 2/9/2020

We also have an example from KIND Karz for attributes, specifically for the airbags. As background, KIND Karz inspectors look for five defect types for each airbag and indicate the number of defects in a log. Again, these are attributes and not continuous measurements, which requires a different analysis approach. Can we rely on the inspection approach? Would the same inspector get the same results twice? What about if two people inspect the same airbag? For this we need to perform an Attribute Agreement Analysis. This is also an important test, as it too is the last line of defense before KIND Karz installs them in their new automobiles.

For this analysis, Albert identified four inspectors with varying degrees of skills and tenure (John, Leslie, Sheila, and Timothy). After Albert completed the analysis, he met again with the project team. "I've good news and bad news," he said. "While overall the inspection process is sufficient in that all the Kappa values are greater than 0.7, the overall accuracy is only 74% and could be as low as 59.66%, so we need to take a look and see what is causing this." See the Minitab output in Fig. 4.10.

"Not to embarrass anyone," he continued. "But we need to understand why one inspector seems to have more variation when repeating an inspection and against a known standard. This likely has something to do with the test instrument, and we'll sort that out in the Analyze phase." After more discussions and taking a closer look at the test equipment, it was determined that the highest incidence of misses was due to one attribute—pinholes. This too will be further evaluated in the Analyze phase, specifically the root cause of the pinhole testing device failing to provide enough distinct category resolution (Fig. 4.11).

Table 4.7 Gage R&R Minitab output

Gage R&R

Variance Components

Source	VarComp	%Contribution (of VarComp)
Total Gage R&R	0.09143	7.76
Repeatability	0.03997	3.39
Reproducibility	0.05146	4.37
Assessor	0.05146	4.37
Part-To-Part	1.08645	92.24
Total Variation	1.17788	100.00

Gage Evaluation

Source	StdDev (SD)	Study Var (6 × SD)	%Study Var (%SV)
Total Gage R&R	0.30237	1.81423	27.86
Repeatability	0.19993	1.19960	18.42
Reproducibility	0.22684	1.36103	20.90
Assessor	0.22684	1.36103	20.90
Part-To-Part	1.04233	6.25396	96.04
Total Variation	1.08530	6.51180	100.00

Number of Distinct Categories = 4

All Appraisers vs Standard

Assessment Agreement

# Inspected	# Matched	Percent	95% CI
50	37	74.00	(59.66, 85.37)

 # Matched: All appraisers' assessments agree with the known standard.

Fleiss' Kappa Statistics

Response	Kappa	SE Kappa	Z	P(vs > 0)
1	0.977897	0.0500000	19.5579	0.0000
2	0.849068	0.0500000	16.9814	0.0000
3	0.814992	0.0500000	16.2998	0.0000
4	0.944580	0.0500000	18.8916	0.0000
5	0.983756	0.0500000	19.6751	0.0000
Overall	0.912082	0.0251705	36.2362	0.0000

Fig. 4.10 Attribute agreement Minitab output

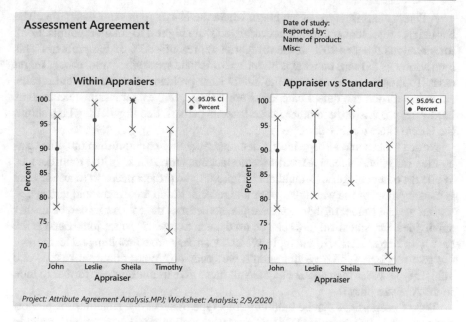

Fig. 4.11 Attribute agreement graph

In conclusion, while there are some concerns to be reviewed further, overall, it seems most of the variation is in the processes and not the measurement systems, so we will continue with the data provided (trusting our measurement systems). As a project team participant or stakeholder, the prior represents the level of detail is normally provided. But as the person performing the analysis, see Sect. 4.4 for detailed instructions and more information as how to perform Gage R&R and Attribute Agreement Analysis.

Now that we know we can generally trust our measurement system, we need to answer the final and perhaps most important question before we wrap up the Measure phase and move to Analyze. That is, are there clues were to look for root causes later? Are there stratification factors?

4.1.5 Funneling and Stratification

Next, we want to funnel our factors and narrow down the potential areas where we might find the root cause when we move to the Analyze phase. First, let's try stratification. This provides clues as to where to focus, and we can leverage graphical tools such as Pareto analysis to visually identify most likely or impactive areas of focus. We can also start to use statistical tools, such as proportion testing.

First let's use Pareto analysis. For this, we will create a Pareto chart using Minitab (Stat/Quality Tools/Pareto Chart). As with prior sections in this chapter, a team member or stakeholder's view will be provided. For the analyst, more detail and how-to is provided in Sect. 4.5 "Pareto Analysis."

A Pareto chart is a type of bar chart, where the X axis represents categories. The bar heights represent counts and are arranged from highest to smallest counts. Look for situations where ~20% of the categories represent ~80% of the problem. This then allows us to focus on what is important or isolate areas where root causes might exist. To illustrate this, consider the KIND Karz problem of warrantee and recalls. As noted earlier, the defect categories are airbags, brakes, emissions, electronics, hardware, propulsion, and other miscellaneous defects. See Fig. 4.12, which shows the Pareto chart for this example.

Notice brakes and airbags have a demonstrably higher count than the other categories (as Albert expected) and together account for 79.1% of the total defects. While the other categories should not be ignored, the project needs to focus on what is causing most of the warrantee work and recalls, which are brakes and airbags.

But we can funnel further. For example, sometimes the categories can be further subdivided. Or, statistical tools can be used, such as test of proportions. Such is the case with KIND Karz. For example, we know airbags are a problem and learn there are two vendors. Which vendor should we focus on? What about brakes? Is one vehicle type worse than another? Could all these give us clues where further to look in the Analyze phase?

Test of proportions can be helpful in this type of scenario. Note: test of proportions can be helpful in other DMAIC phases as well but will be shown here and in Sect. 4.6 to illustrate specifically how to use and apply these tools. As with prior sections, the content as follows will be limited to what is typically shared with project teams and stakeholders.

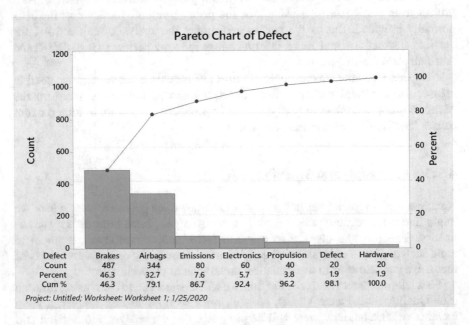

Pareto Chart of Defect

Defect	Brakes	Airbags	Emissions	Electronics	Propulsion	Defect	Hardware
Count	487	344	80	60	40	20	20
Percent	46.3	32.7	7.6	5.7	3.8	1.9	1.9
Cum %	46.3	79.1	86.7	92.4	96.2	98.1	100.0

Project: Untitled; Worksheet: Worksheet 1; 1/25/2020

Fig. 4.12 Pareto chart, KIND Karz defect categories

Table 4.8 Airbag vendor defect rates

Category	Vendor 1	Vendor 2
Airbags with issues	200	180
Total airbags used	50,000	40,000

Fig. 4.13 Test of two
proportions Minitab output

Test and CI for Two Proportions

Method

p_1: proportion where Sample 1 = Event
p_2: proportion where Sample 2 = Event
Difference: $p_1 - p_2$

Descriptive Statistics

Sample	N	Event	Sample p
Sample 1	50000	200	0.004000
Sample 2	40000	180	0.004500

Estimation for Difference

Difference	95% CI for Difference
-0.0005	(-0.001358, 0.000358)

CI based on normal approximation

Test

Null hypothesis	H_0: $p_1 - p_2 = 0$
Alternative hypothesis	H_1: $p_1 - p_2 \neq 0$

Method	Z-Value	P-Value
Normal approximation	-1.14	0.253
Fisher's exact		0.255

Let's review the first example for airbags. To set up the problem, determine how many defects we are seeing at each vendor. This is shown in Table 4.8.

Here the test of two proportions can help answer whether one vendor is worse than the other. If so, the project can focus on the worst vendor in the Analyze phase (again, this is another way to funnel). After completing the analysis, Albert again reviewed the results with the team (see Fig. 4.13). Because the P-value is greater than 0.05, he could not conclude one vendor was different than the other. However, that could be because there is excessive variation, or the sample size is too small. Either way, the team will need to visit and evaluate defects at both vendors in the Analyze phase.

Next, Albert considered the brake issue. Could one vehicle be the worse than another? The problem is set up as shown in Table 4.9, and the analysis method will be Chi-Sq test of multiple proportions.

Table 4.9 Chi-sq table for brake defect counts by vehicle

	Vehicle 1	Vehicle 2	Vehicle 3	Vehicle 4	Vehicle 5
Brake warrantee issues	200	20	50	50	0
No brake warrantee issues	749,800	649,980	249,950	49,950	450,000
Total vehicles produced	750,000	650,000	250,000	50,000	450,000
Percent defective	0.027%	0.003%	0.020%	0.100%	0.000%

Tabulated Statistics: Category, Worksheet columns

```
Rows: Category   Columns: Worksheet columns

                      Vehicle 1   Vehicle 2   Vehicle 3   Vehicle 4   Vehicle 5        All

Brake Warrantee Issues       200          20          50          50           0        320
                            0.03        0.00        0.02        0.10        0.00       0.01
                           111.6        96.7        37.2         7.4        67.0
                         69.9612     60.8788      4.3968    243.3794     66.9767
                                                              Greatest contribution to ChiSq
No Brake Warrantee Issues 749800      649980      249950       49950      450000    2149680
                           99.97      100.00       99.98       99.90      100.00      99.99
                        749888.4    649903.3    249962.8     49992.6    449933.0
                          0.0104      0.0091      0.0007      0.0362      0.0100

All                       750000      650000      250000       50000      450000    2150000
                          100.00      100.00      100.00      100.00      100.00     100.00

Cell Contents
   Count
   % of Column
   Expected count
   Contribution to Chi-square
```

Chi-Square Test

	Chi-Square	DF	P-Value
Pearson	445.659	4	0.000
Likelihood Ratio	390.308	4	0.000

Choose smaller of the following P-values:
P-value for larger sample sizes
Tends to be more accurate for smaller sample sizes

Fig. 4.14 Minitab Chi-sq results

After completing the analysis, Albert shared the results with the team. "This analysis is a bit more challenging to explain," he said. "But let's go over the results." He shared Fig. 4.14. Given the overall P-value is low (less than 0.05), we can conclude at least one proportion is different than the others. As shown, Vehicle 4 has the greatest contribution to Chi-sq, the distribution used to perform the analysis. So, in the Analyze phase, we will also take a close look at Vehicle 4 to see why we are having more recalls for that automobile type. Again, what has been shared so far are the outcomes of the analysis. See Sect. 4.6 "Test of Proportions" for more details and step-by-step instructions on how to do and more fully interpret the results.

This concludes the Measure phase. We have a sense of the process stability and capability, we generally can trust our data, and we have ideas where to review further for root causes in the Analyze phase. But before we transition to the Analyze phase, lets summarize our findings.

Summarizing at the end of each phase is important to ensure everyone has a common understanding as to the outcomes of each DMAIC phase. For the Measure phase, we need to list the funneled areas to be carried forward to the Analyze phase, where we will finally get to root cause. As a general good practice, don't make the reader work at it to understand your storyboard. Don't write to impress, and tailor the findings such that they are comprehensible to a reasonably informed audience. There should be a clear story and thread throughout the storyboard. So, for KIND Karz, we have concluded the following in the Measure phase:

1. Processes are not capable for the overall defect rate for airbags, airbag fabric tensile strength, and brake torsion resistance.
2. Airbags and brakes have the highest incident counts for recalls and warrantee claims in the first 12 months after a new KIND Karz purchase.
3. No difference between vendors detected for airbag defects.
4. Vehicle 4 tends to have more brake defects than the other vehicles.
5. While the airbag test method was shown to be adequate as designed, it yielded low accuracy for the pinhole test.
6. Airbag defects worsened after the new assembler was installed at the vendor.

Now let's proceed to Analyze to determine root causes. But first you might want to review the details of the statistical methods, which are provided in the sections immediately following this section.

4.2   Control Charts: Minitab Methods and Analysis Detail

4.2.1   Common Cause Versus Special Cause Variation

To begin, consider the differences between common and special cause variation. Common cause variation is just the normal random variation in a process. Some data points might be higher than the mean, but some are lower but do not indicate something unusual has happened to the process. However, special cause variation does indicate something unusual has happened.

For example, think about what time you arrive at work. If your start time is 8:00 AM, perhaps you normally arrive between 7:45 and 7:55 AM, even if you leave home about the same time. That is due to normal variation in traffic, variability in speed, timing of traffic lights, and a myriad of other things. Imagine your boss gets really upset one day when he notices that you arrived at 7:55 AM and demands that you leave earlier. So, you start leaving 15 min earlier and notice you now arrive between 7:30 and 7:40 AM but with the same variation in time. This is common cause variation—to change it, a fundamental change to the process would be required.

But let's continue with this example. With your originally leave time, 1 day you arrive at 8:15 AM. What could cause that? There are many things—perhaps a car accident in front of you, or you have a flat tire, or your child is sick. That is an example of a special cause event occurring – something unusual happened. So, your boss shouts at you again, and demands you leave earlier. In this case, there isn't a fundamental problem with your route or time to leave, but something unusual has happened. Changing the process (or leave time) in this case isn't the solution.

In practice, common cause variation is what you should expect when the process is performing normally. You might not like that variation, but you will need to fundamentally change the process to affect it. However, special cause variation is something out of the ordinary—a signal that something unusual has happened. You should investigate and resolve the special cause event and don't treat it like an ordinary event. Understanding which (special or common) early on dictates how on should approach problem solving. Using the wrong approach can make the problem worse.

When we detect special cause signals, that indicates an unstable process. Control charts are used to help us determine whether our process is stable and if there are special cause events occurring.

4.2.2 Control Chart Fundamentals

A control chart is essentially a time series chart with three lines added: a lower control limit (LCL), the mean or average, and an upper control limit (UCL). (A time series chart shows what is being measured on the Y axis, in chronological order on the X axis from left to right.)

The UCL and LCL's are approximately +/−3 sigma from the mean and account for 99.7% of the variation (for control limits, standard deviation is calculated differently than traditional descriptive statistics). Any data points outside the control limits are not the norm and represent a special cause event. Also, non-random patterns in the control chart are considered special cause as well. A control chart example is shown in Fig. 4.15. There are two points flagged in red as being special cause. Due to this, the process cannot be stable.

Again, special cause signals represent unusual data points. There are a variety of tests used to determine if such a situation occurs. The more common ones are related to whether points are outside the control limits, or indicate a shift in the mean, or represent a trend or an alternating pattern. See Fig. 4.16 for an example from Minitab for an I-MR chart (one of many charts available in Minitab but perhaps the most used). Notice the first four are selected. For an I-MR chart, the first four do not require normal data. However, the last four do require normally distributed data. Note: However, it has been the author's experience when there are different distributions that are not normal, false alarms can occur for special causes (with highly shifted distributions).

In most cases and in practice, only the first four tests need to be assigned. (Note for the second test, the value shown is 8 versus Minitab's default 9, to provide a quicker signal for process shifts.)

Minitab offers a variety of charts, all of which should be considered based on your specific application. However, this book will focus on four chart types that are most common in practice—the I-MR, Xbar-R, P, and U-charts. We will also review the Run Chart, a related and helpful tool.

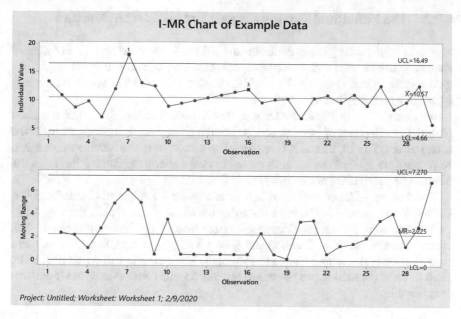

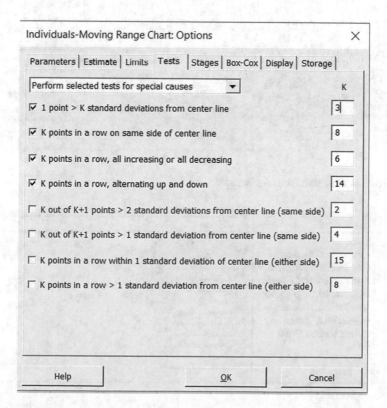

Fig. 4.15 Control chart example

Fig. 4.16 Minitab example: special cause tests

4.2.3 The Individual Moving Range Control Chart in Minitab

The individual moving range control chart (I-MR) is the most versatile of control charts and can be used with most data types. However, other chart types might provide quicker special cause signals (will be introduced later) when applied to the data types for which they are designed.

An example of an I-MR charter was shown prior, see Fig. 4.15. The top chart plots the actual values of interest, and the bottom chart plots the moving range of the actual values. The UCL and LCL are based on outcomes of the moving range chart, so if there are special causes in the moving range chart, the control limits on the individuals chart might not be reliable.

To create an I-MR control chart in Minitab, select Stat/Control Charts/Variables Charts for individuals/I-MR. For a visualization of the Minitab path, see Fig. 4.17.

The following job aid provides step-by-step instructions how to create an I-MR control chart in Minitab. Recall in the KIND Karz case study, there was concern regarding the brake caliper torsion rod strength. This example will develop an I-MR chart to see if the torsion rod resistance has any special cause events and if the process is stable.

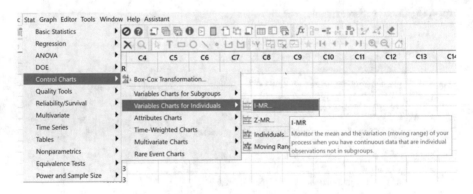

Fig. 4.17 Minitab path for an I-MR chart

Step	Instructions	Illustration
1	The data • Paste in the example data from the available data file (control Charts tab). Partial table shown here for illustration. • Select the "brake caliper torsion IMR" data.	A 2 Brake Caliper Torsion IMR 3 76.66 4 80.37 5 77.25 6 79.07 7 77.33 8 76.98 9 78.23 10 78.96 11 76.92 12 78.88 13 77.69

Step	Instructions	Illustration
2	Creating the I-MR • Select stat/control Charts/variables Charts for individuals/I-MR. • For "variables," select "brake caliper torsion IMR". • Then select "ok".	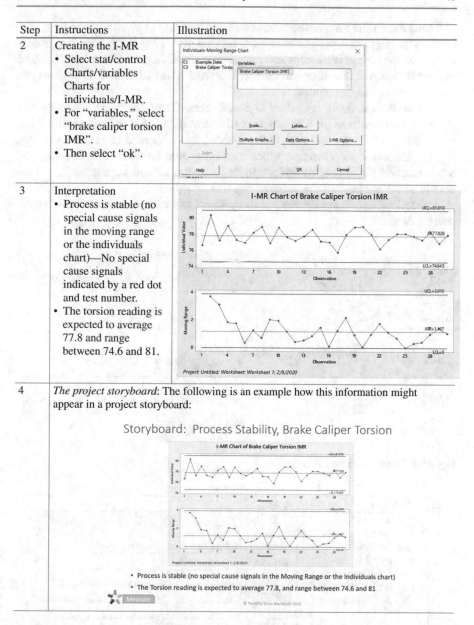
3	Interpretation • Process is stable (no special cause signals in the moving range or the individuals chart)—No special cause signals indicated by a red dot and test number. • The torsion reading is expected to average 77.8 and range between 74.6 and 81.	
4	*The project storyboard*: The following is an example how this information might appear in a project storyboard:	

Storyboard: Process Stability, Brake Caliper Torsion

* Process is stable (no special cause signals in the Moving Range or the Individuals chart)
* The Torsion reading is expected to average 77.8, and range between 74.6 and 81

4.2.4   The P-Chart for Proportional Data in Minitab

While the I-MR chart can be used for a variety of data types, it is especially intended for continuous data. However, when other data types are used, other charts are better suited. Such an example is for proportional data, where the P-chart is recommended. Proportions are common for such things as yield or scrap.

Using the KIND Karz case study, remember the CTQ or Y is the number of warrantee or recall claims per 1000 automobiles sold after 12 months. This is a proportion, or on average 200/1000, or 200 recalls or warrantees per 1000 vehicles or 20% of the vehicles had recalls or warrantees. Use the P-chart when both defectives and non-defectives can be counted.

The Minitab path for a P-chart is Stat/Control Charts/Attribute Charts/P. See Fig. 4.18 for a visual of the Minitab path. Also note there are two options—the Lane P and the P. While we will review the P-chart, it is also recommended to run the P-chart diagnostic to determine which method should be used. This is especially recommended if the control limits appear to be too narrow or too wide, but the details of this are beyond the scope of this text.

The following job aid provides step-by-step instructions how to create P-chart chart in Minitab.

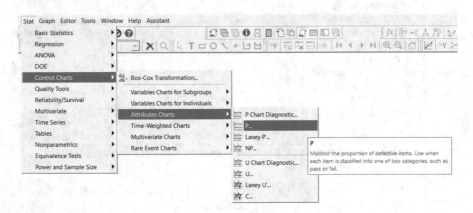

Fig. 4.18 Minitab path for a P-chart

Step	Instructions	Illustration		
1	The data • Paste in the example data from the available data file (control Charts tab). • Partial table shown here for illustration.	**C**	**D**	**E**	
		Month	Cars Produced	WarRec per 1000 PChart	
		Jan 2018	1013	192	
		Feb 2018	992	195	
		Mar 2018	960	191	
		Apr 2018	996	207	
		May 2018	983	194	
		Jun 2018	955	195	
		Jul 2018	951	195	
		Aug 2018	984	204	
		Sep 2018	1023	206	
		Oct 2018	986	209	
		Nov 2018	967	198	
		Dec 2018	1017	194	

Step	Instructions	Illustration
2	Creating the P-chart • Select stat/control Charts/attribute Charts/P. (note: If the data pasted as text into Minitab, right click in the column and select "format column" and change to "automatic numeric"). • For "variables," select "WarRec per 1000 P chart." this is the numerator of the proportion. • For "subgroup sizes" select "cars produced." this is the denominator of the proportion. • Then select "scale".	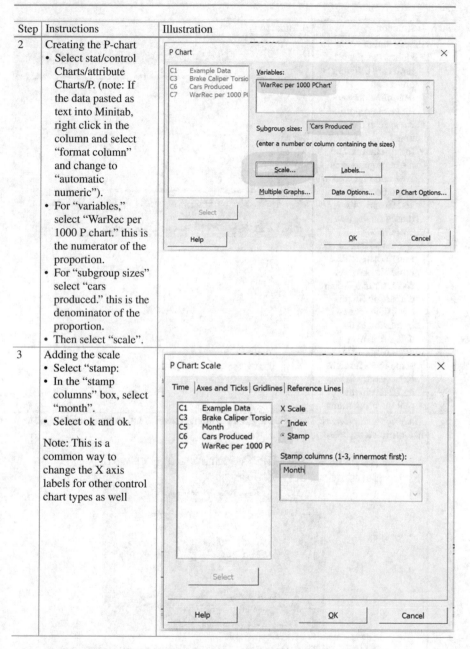
3	Adding the scale • Select "stamp: • In the "stamp columns" box, select "month". • Select ok and ok. Note: This is a common way to change the X axis labels for other control chart types as well	

Step	Instructions	Illustration
4	Interpretation • Notice the control limits are different given the denominators vary for each proportion. • No special cause noted. • Interpretation: The mean proportion is 20% and expected to range between 16.5% and 24% (based on the last proportion). • For the advanced user: When control limits do not seem correct, run a P-chart diagnostic (in this case, they seem wide in relation to the data). Run a diagnostic, or in some cases it might be necessary to use an I-MR chart to plot the proportions.	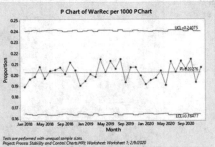

| 5 | *The project storyboard*: The following is an example how this information might appear in a project storyboard: |

Storyboard: Y (CTQ): Proportions of Cars with Warrantee Claims or Recalls

• No special cause noted – process is stable

• Interpretation: The mean proportion is 20%, and expected to range between 16.4% and 24%

4.2.5 The X-Bar R Chart for Systematically Sampled Data in Minitab

A control chart that is recommended in lieu of the I-MR chart when the data is gathered in subgroups is the X-Bar R chart, which is for data that is in subgroups. This will allow us to also understand within-subgroup variation, which the I-MR would not do if we just plotted the average subgroup values.

Use the X-Bar R chart when data is sampled in subgroups between 2 and 8 samples. When subgroup sizes are 9 or greater, use the X-Bar S chart.

Subgroup data is gathered systematically. For example, let's return to the KIND Karz case study and the production of brake calipers. On the line that finishes the calipers after molding, three calipers are pulled every 30 min for torsion resistance. This is the data we want to use in a control chart to determine whether our process is stable.

The Minitab path for an X-Bar R Chart is Stat/Control Charts/Variables Charts for Subgroups/Xbar-R. See Fig. 4.19 for a visual of the Minitab path.

The following job aid provides step-by-step instructions how to create an Xbar-R chart in Minitab.

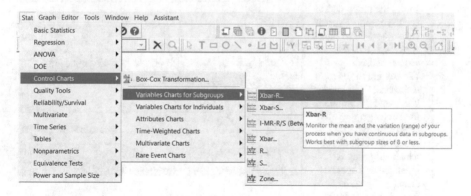

Fig. 4.19 Minitab path for the Xbar-R cart

Step	Instructions	Illustration	
1	The data • Paste in the example data from the available data file (control Charts tab). Partial table shown here for illustration. • Select the "XBarR caliper torsion" data. • Notice the structure of the data—Subgroups are in order.	**G** Subgroup	**H** XBarR Caliper Torsion
		1	76
		1	77
		1	78
		2	77
		2	80
		2	79
		3	80
		3	77
		3	80
		4	80

Step	Instructions	Illustration
2	Creating the P-chart • Select stat/control Charts/variables Charts for subgroups/Xbar-R. • For "variables," select "XBarR caliper torsion". • For "subgroup size" select the column that associates the data point with the subgroup (for this example "subgroup.") if all subgroup sizes are the same and, in the order, shown prior, the subgroup numeric size can be included instead. • Select "ok".	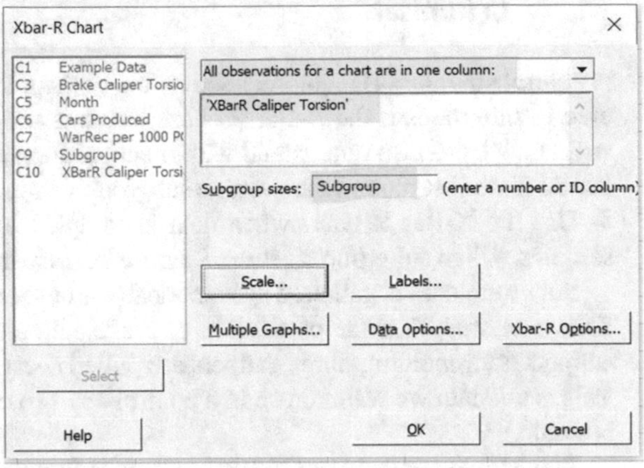
3	Interpretation • The bottom graph shows the range within each subgroup. Special causes here show which subgroup (10) has special cause. • The top graph shows the average within a group and the variation then between subgroups. Special cause here (10) shows which subgroup is indicated as a special cause. • Expect the average of any subgroup to be 77.43 and between 73.4 and 81.5. However, this might not be reliable given the special cause signals noted, especially in the moving range chart from which the control limits are determined. • Need to investigate what was different with subgroup 10.	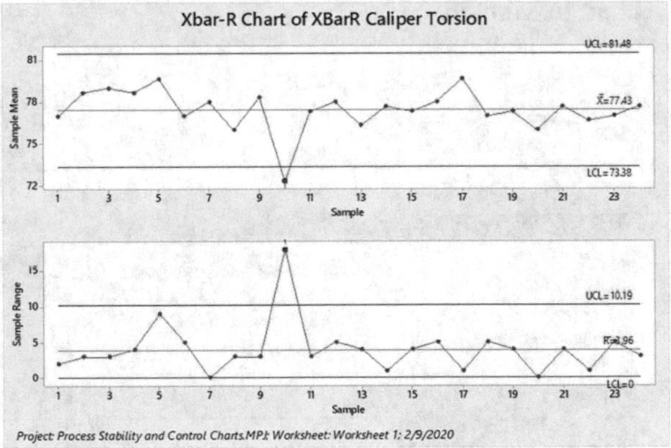 Xbar-R Chart of XBarR Caliper Torsion Project: Process Stability and Control Charts.MPI: Worksheet: Worksheet 1: 2/9/2020

Step	Instructions	Illustration
4	*The project storyboard*: The following is an example how this information might appear in a project storyboard:	

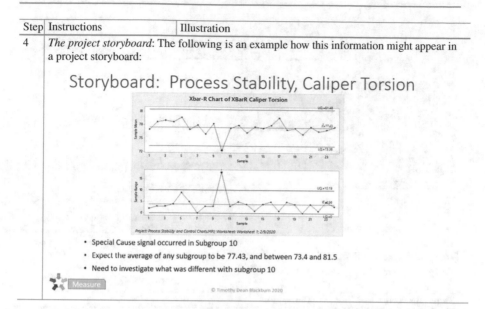

- Special Cause signal occurred in Subgroup 10
- Expect the average of any subgroup to be 77.43, and between 73.4 and 81.5
- Need to investigate what was different with subgroup 10

4.2.6   Control Chart for Attributes: The U-Chart in Minitab

In some situations, defects can be counted, but non-defects cannot be counted. For example, suppose you are an inspector for a paint shop. One of your jobs is to count the number of scratches in the paint of all cars painted in a day. You could do this, but you would not be able to count how many scratches aren't there.

But there often is still a difference in opportunity that should be considered (e.g., number of cars produced, days in the month, etc.). For example, you would expect to see more scratches when 100 cars are painted in a day versus 80. In this case, use the U-chart, which is for unequal opportunities. If opportunities were the same, use the C-chart.

Like the comments for P-chart, if the control limits appear to be too wide or too narrow, run a U-chart diagnostic. In such a case it might be necessary to run Laney's U-chart.

To illustrate how to create a U-chart, consider the number of defects detected in airbag tests for the KIND Karz example. While we can inspect and observe a variety of defects, we can't count the number of no defects. For example, if we are counting how many pinholes are in the fabric, we can't count how many pinholes are not there (that is nonsensical).

The Minitab path for a U-chart is Stat/Control Charts/Attributes Charts/U. See Fig. 4.20 for a visual of the Minitab path.

The following job aid provides step-by-step instructions on how to create a U-chart in Minitab.

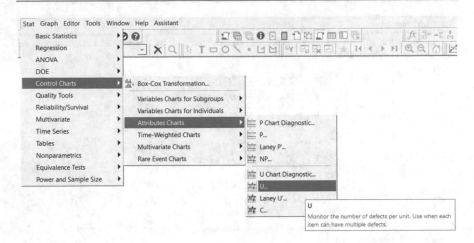

Fig. 4.20   Minitab path for a U-chart

Step	Instructions	Illustration
1	The data • Paste in the example data from the available data file (control Charts tab). Partial table shown here for illustration. • Select the "U-chart" data. • Notice in the data that the number of airbags produced in a month could vary, so there is an unequal opportunity, in this case, the U-chart is the correct option versus the C chart.	<table><tr><th>K</th><th>L</th><th>M</th><th>N</th></tr><tr><td>Month_1</td><td>Airbags Produced</td><td>Defects Identified</td><td>Shift</td></tr><tr><td>Jan 2018</td><td>1013</td><td>192</td><td>Before</td></tr><tr><td>Feb 2018</td><td>992</td><td>195</td><td>Before</td></tr><tr><td>Mar 2018</td><td>960</td><td>100</td><td>Before</td></tr><tr><td>Apr 2018</td><td>996</td><td>207</td><td>Before</td></tr><tr><td>May 2018</td><td>983</td><td>194</td><td>Before</td></tr><tr><td>Jun 2018</td><td>955</td><td>195</td><td>Before</td></tr><tr><td>Jul 2018</td><td>951</td><td>195</td><td>Before</td></tr><tr><td>Aug 2018</td><td>984</td><td>204</td><td>Before</td></tr><tr><td>Sep 2018</td><td>1023</td><td>206</td><td>Before</td></tr><tr><td>Oct 2018</td><td>986</td><td>209</td><td>Before</td></tr><tr><td>Nov 2018</td><td>967</td><td>198</td><td>Before</td></tr><tr><td>Dec 2018</td><td>1017</td><td>194</td><td>Before</td></tr></table>
2	Creating the U-chart • Select stat/control Charts/attributes Charts/U. • For "variables," select "defects identified". • For "subgroup sizes," select "airbags produced". • Select "scale".	U-Chart dialog box: Variables: 'Defects Identified'; Subgroup sizes: 'Airbags Produced' (enter a number or column containing the sizes). Buttons: Scale..., Labels..., Multiple Graphs..., Data Options..., U Chart Options..., Select, Help, OK, Cancel. Column list: C1 Example Data, C3 Brake Caliper Torsio, C6 Cars Produced, C7 WarRec per 1000 P, C9 Subgroup, C10 XBarR Caliper Torsi, C13 Airbags Produced, C14 Defects Identified.

Step	Instructions	Illustration
3	Scale • Select "stamp". • In the "stamp columns" box, select "Month_1". • Click ok and ok.	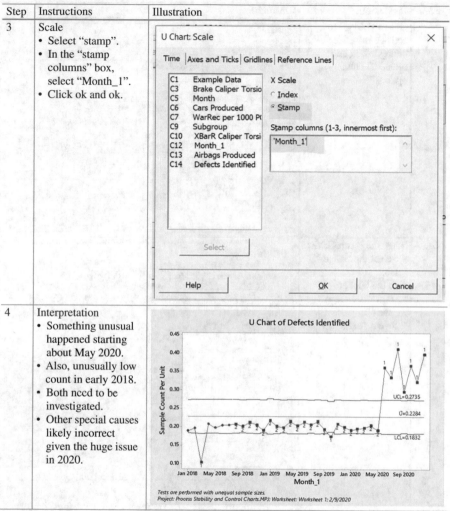
4	Interpretation • Something unusual happened starting about May 2020. • Also, unusually low count in early 2018. • Both need to be investigated. • Other special causes likely incorrect given the huge issue in 2020.	

Explanation:

After further discussions, Albert learned that a new airbag assembler machine was installed June 2020. Also, he discovered that the low data value occurred in March 2018 and was in error—the plant was shut down that month and the Quality Data Base default was entered automatically. So next he reran the control chart to show a shift occurring and removed the point that he knew was incorrect. This is important as we would like to understand what the process was before and after the shift

Step	Instructions	Illustration
5	Hold ctrl and E to activate the prior dialogue box and select "U chart options"	
6	Stage the control chart • Select the "stages" tab. • In the "define stages" box select "shift". • Select "ok".	
7	Removing the outlier • Select "data options".	

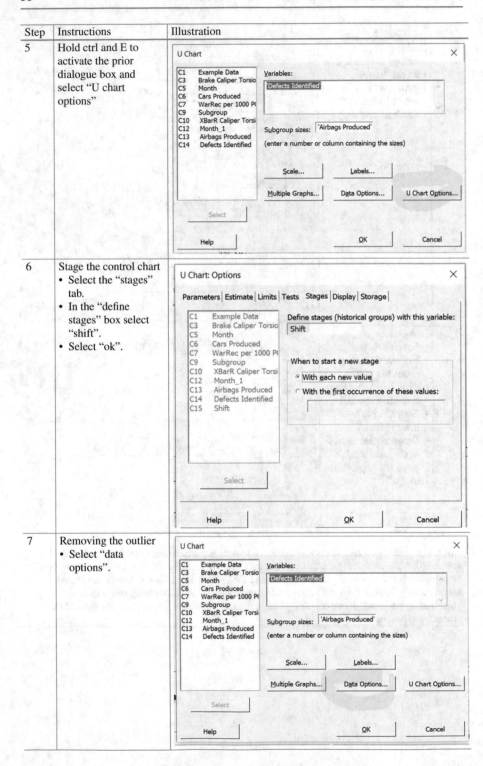

Step	Instructions	Illustration
8	Removing the outlier (continued) • Select "specify which rows to exclude". • Select "row numbers". • Enter 3 in the box, which is the data point determined to be in error. • Select ok and ok.	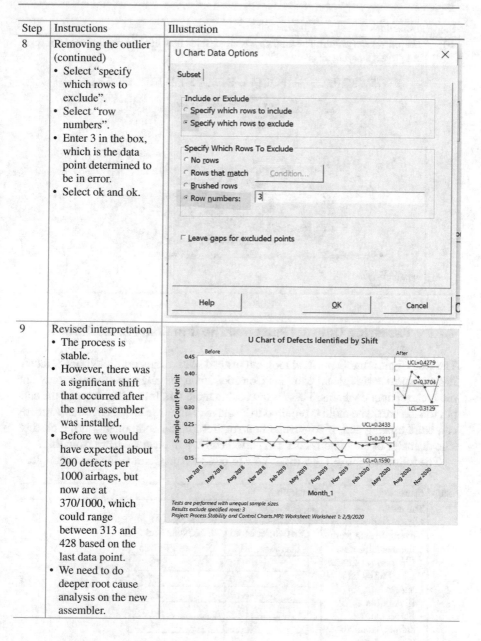
9	Revised interpretation • The process is stable. • However, there was a significant shift that occurred after the new assembler was installed. • Before we would have expected about 200 defects per 1000 airbags, but now are at 370/1000, which could range between 313 and 428 based on the last data point. • We need to do deeper root cause analysis on the new assembler.	

Step	Instructions	Illustration
10	*The project storyboard*: The following is an example how this information might appear in a project storyboard:	

Storyboard: Airbag Defects Process Stability

- Process stable but shifted after the new assembler was installed

© Timothy Dean Blackburn 2020

4.2.7   Assessing Unusual Patterns: The Run Chart in Minitab

The Run Chart is another useful tool, but instead of ascribing control limits, it identifies statistically significant patterns. Consider again the KIND Karz case study and the brake caliper concerns. There is a reusable mold that forms the shape of the caliper torsion part. The mold is required to be cleaned after use and if not done correctly can build up slag and, if so, results in a smaller torsion component during casting. The diameter of the mold is checked periodically. What can we learn from this?

The following job aid illustrates how to build a Run Chart using this example.

Step	Instructions	Illustration	
1	The data	**P**	**Q**
	• Paste in the example data from the available data file (control Charts tab). Partial table shown here for illustration.	Ex Caliper Mold Diameter IMR	Ex Caliper Mold Diameter RunCh
		19.50	19.50
		20.90	20.90
		21.20	21.20
		21.36	21.36
	• Select the "ex caliper Mold" data.	22.22	22.22
	• There are two columns of data. The first will be used for an I-MR chart, and the second is a subset of the data, for which Albert is concerned there might be a trend occurring.	23.41	23.41
		24.10	24.10
		24.80	24.80
		25.20	25.20
		25.80	25.80
		23.31	
		24.10	

Step	Instructions	Illustration

2	The I-MR chart and interpretation
	• See prior instructions on how to create an I-MR chart but create an I-MR chart using the first column of data.
	• Several questions come to mind:
	  ◦ What are the special causes here?
	  ◦ Could it be things were shifting up then remained at the new level?
	  ◦ How can we tell?
	• Next Albert evaluated a subset of the early data which the team suspected was a trend leading to a shift in the mean and used Minitab to generate a run chart.
		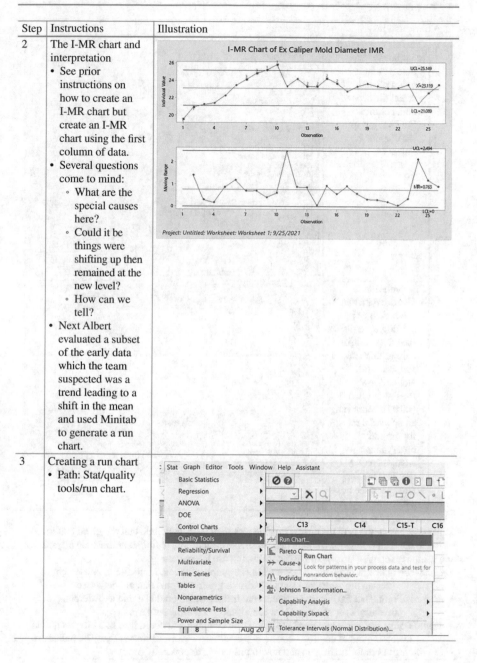

3	Creating a run chart • Path: Stat/quality tools/run chart.	

Step	Instructions	Illustration
4	Creating a run chart (continued) • Select "single column" and select the second data column. • For "subgroup size," enter 1. • Select ok.	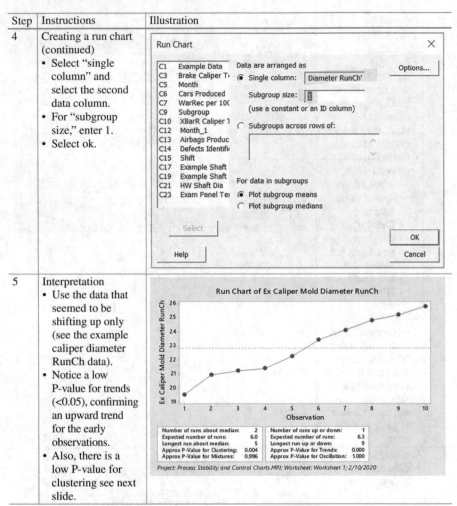
5	Interpretation • Use the data that seemed to be shifting up only (see the example caliper diameter RunCh data). • Notice a low P-value for trends (<0.05), confirming an upward trend for the early observations. • Also, there is a low P-value for clustering see next slide.	

Understanding the patterns:
Whenever any of the patterns have a P-value less than 0.05, conclude that the given pattern exists in the data (reference Minitab help for these definitions) (Minitab online help https://www.minitab.com/en-us/support/). The following are the patterns:
• *Clustering*: Groups of points in one area of the graph, and could indicate measurement or set-up problems, or sampling from a group of items that are defective.
• *Mixture*: Points that frequently cross the center line, which could be due to different populations, or two processes functioning at a different level.
• *Trend*: Points indicate a sustained drift either up or down and could lead to a process quickly becoming out of control. Change in operators, or worn equipment may be indicated.
• *Oscillating*: Data fluctuating up and down, indicating an unsteady process.

Step	Instructions	Illustration
6	*The project storyboard*: The following is an example how this information might appear in a project	

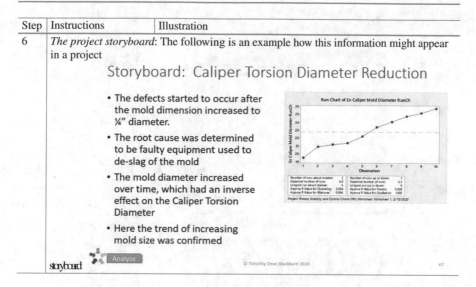

4.3   Capability Analysis: Minitab Methods and Analysis Detail

4.3.1   Introduction to Capability Analysis

The voice of the process (e.g., UCL, LCL) was considered in the sections on control charts. However, a question remains, "Is the process capable of meeting customer needs?" That is, will the process operate within the Upper and Lower Specification Limits (USL, LSL)—these are set by the customer, not the process. It is important to know in the Measure phase whether our process is capable of meeting customer needs. Also, in the Control phase, the capability of the process needs to be revisited. The methods reviewed in these sections are helpful for both phases.

There are three primary possibilities when considering whether the process is capable. The first is when the UCL and LCL exactly align with the USL and LSL as shown in Fig. 4.21. If this were to occur, the process would be considered *just capable* or *marginally capable*. Later we will review how a quantitative process capability value will be calculated, or Ppk. In this case, the Ppk would be 1.

The next example as shown in Fig. 4.22 is for a capable process where the control limits fit well within the specification limits. Such a process would have a Ppk equal to or greater than 1, with a goal of 1.33 or higher in most cases.

The final example is when one or both control limits are worse than a specification limit. In this case, the Ppk will be less than 1, and the process will be considered not to be capable (Fig. 4.23).

Specification limits (USL, LSL) should not be shown on control charts. People might confuse specification limits with control limits and lead to incorrect actions. For example, consider a process that is stable but has a low Ppk. An operator might try to implement special cause solutions to chase a data point and make the process

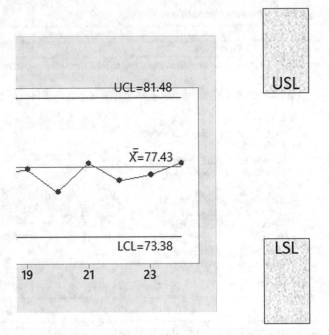

Fig. 4.21 Just capable process

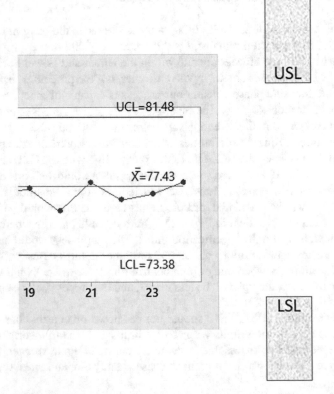

Fig. 4.22 Capable process

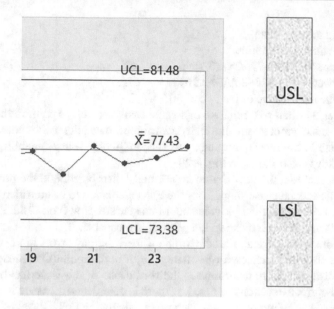

Fig. 4.23 Process not capable

mean or average worse. Or consider a process not stable but within spec limits. This could give the operator a false sense that everything is acceptable, while the process continues to drift out of spec.

There are two measures for process capability typically used in industry. The first is Cpk, which is for shorter term process capability (or within batch variability). The other is Ppk, which is for longer term process capability (or between batch variability) and the recommendation of this book as the primary index for capability. The calculations are similar, except the standard deviation is calculated differently. For Ppk, the standard deviation is calculated the same as for descriptive statistics, whereas for Cpk it is a pooled standard deviation when replicates are two or more (i.e., samples are two or more). Minitab uses another method to estimate standard deviation for Cpk from control charts when it is one replicate (i.e., sample sets are one) (Minitab Online Help https://www.minitab.com/en-us/support/).

To illustrate how to calculate Ppk, consider an example from KIND Karz. Remember we want the caliper torsion rods to have at least 80 N-m of resistance. In the last section, a control chart was developed for this. The mean torsion resistance was 77.828, and the standard deviation was 0.964.

Use the following equation to calculate Ppk. Note there is three sigma in the denominator. That is because most values (99.73%) are expected to fall within three sigma of the mean and then is expected to account for nearly all the random variation in a process. So, if the difference between the mean and the actual value is equal to three times sigma, the Ppk will be 1. The USL = 100 and LSL = 80.

$$Ppk = \min\left[\frac{USL - \bar{X}}{3\sigma} \ or \ \frac{\bar{X} - LSL}{3\sigma}\right]$$

where

- $\bar{X}$ = the sample mean
- σ = the standard deviation
- Ppk (upper) = (100–77.828)/(3*0.9642) = 7.67.
- Ppk (lower) = (77.828–80)/(3*0.9642) = −0.75.
- Ppk = the lower value of the two = −0.75.

However, Minitab will generate this value easily as well as confirm the assumption of a stable process and normally distributed data. For continuous data, the Minitab path to run a capability analysis is Stat/Quality Tools/Capability Sixpack/Normal. This is also shown in Fig. 4.24.

For this analysis, the data should be normally distributed, and the process must be stable (no special cause signals) for the Ppk calculation to be considered reliable. If the data is unstable, the Ppk value could inaccurate. However, if the Ppk is very low, it still can give an indication the process is incapable.

If the data is not normal, first determine if there are any errors in the data. If so, correct the error and recheck for normality. Or, if outlier point(s) are leading to the non-normal distribution, determine if the root cause of the issue has been determined and corrective actions initiated. If there is confidence the issue is unlikely to repeat, it also can be removed from purpose of calculating Ppk. However, note these in your analysis report or storyboard. If the points cannot be removed, first determine if an alternative distribution is a better fit. In some cases, a data transformation could also be needed, or other nonparametric methods used (beyond the scope of this book).

Sample size is also important. The data should represent the full expected variation of the process. Smaller sample sizes result in larger confidence intervals and less confidence as to the true Ppk value. See Table 4.10 which shows the upper CI values for Ppk for various sample sizes. When using smaller sample sizes, it is recommended the minimum target values be as shown to have confidence the true Ppk is 1 or higher.

To illustrate how to perform this analysis, a step-by-step job aid is provided in the following, based on the same prior example for the brake calipers. Later, another job aid is provided for non-normal data and later for discrete data.

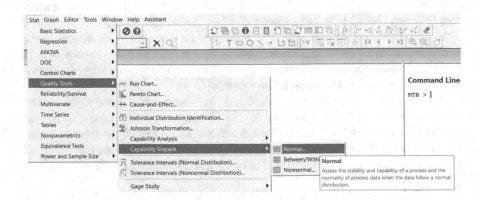

Fig. 4.24 Minitab path for capability analysis

Table 4.10   Upper CI for Ppk

Sample size	Upper CI for Ppk = 1
25	1.42
50	1.27
100	1.18

4.3.2   Using Minitab to Calculate Ppk for Continuous Data

Step	Instructions	Illustration
1	The data • Paste in the example data from the available data file (process capability tab). Partial table shown here for illustration. • Use the "brake caliper torsion IMR" data. (this is the same data used in the prior section to develop the control chart).	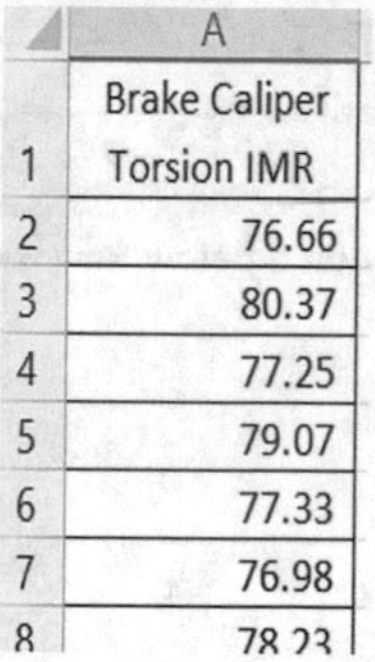 A / Brake Caliper Torsion IMR / 2 76.66 / 3 80.37 / 4 77.25 / 5 79.07 / 6 77.33 / 7 76.98 / 8 78.23
2	Ppk analysis in Minitab • Stat/quality tools/ capability Sixpack/ Normal. • Select "single column" and choose the "brake caliper torsion IMR" data. • In "subgroup size," enter 1. • For the lower spec, enter 80 (the minimum torsion resistance KIND Karz identified as being safe). • For the upper spec, enter 100 (too much torsion resistance in this case is not desired and can lead to breaking a linkage arm). • Select ok.	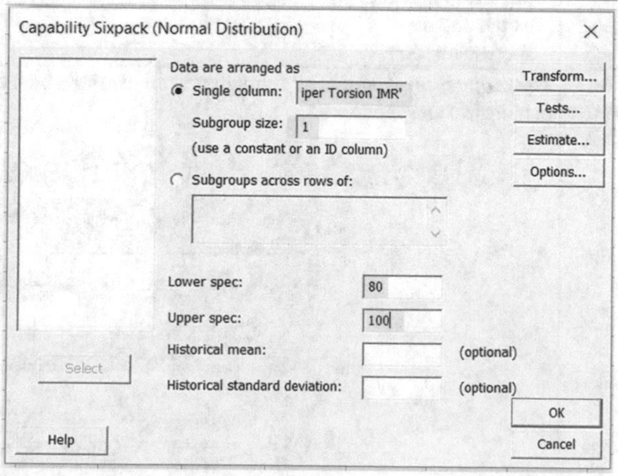 Capability Sixpack (Normal Distribution) — Data are arranged as: Single column: 'iper Torsion IMR'; Subgroup size: 1 (use a constant or an ID column); Subgroups across rows of: ; Lower spec: 80; Upper spec: 100; Historical mean: (optional); Historical standard deviation: (optional); Transform... Tests... Estimate... Options... Select / Help / OK / Cancel

Step	Instructions	Illustration
3	Interpreting the results • Ppk is −0.75, <1, process is not capable. ◦ Note this matches the earlier manual calculation. ◦ It is negative since the mean is less than the LSL (see the prior equation). • Check assumptions. ◦ Normality: P-value >0.05 so the assumption of normality is met. ◦ Process (I-chart) is stable. • If either of the assumptions above were not met, the Ppk estimate might not be reliable. See prior for guidance as what to do in such a situation.	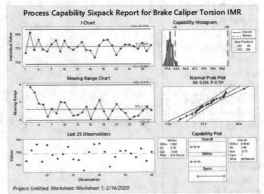 See the storyboard below for a larger image.
4	*The project storyboard*: The following is an example how this information might appear in a project storyboard:

Storyboard: Ppk for Brake Calipers

Ppk is -0.75, <1, process is not capable

4.3.3 Using Minitab to Calculate Ppk for Continuous Data: Non-normal

In some cases, the data might not be normally distributed. After reviewing the data points leading to non-normality, it might be concluded the points must be retained. The next step then is to determine if the dataset follows a distribution different than normal.

While somewhat an advanced topic for this text, first go to Stat/Quality Tools/ Individual Distribution Identification and determine if the data fits another distribution. If so, then use the non-normal distribution tool in Minitab and Stat/Quality Tools/Capability Sixpack/Nonnormal and choose the non-normal distribution. But if the data does not follow another distribution, and is typically expected to be normally distributed, a data transformation can be considered. But it is recommended the analyst consult with a statistician or subject matter expert until expertise is gained prior to transforming.

But in some cases, data transformation can be helpful and can be used to more accurately calculate Ppk. Such a situation exists in the KIND Karz case. Albert wishes to calculate Ppk for the Y or CQA tensile strength for airbags.

Step	Instructions	Illustration
1	The data • Paste in the example data from the available data file (process capability tab). Partial table shown here for illustration. • Select the "airbag tensile strength" data.	**C** Airbag tensile strength 40.86 13.82 3.92 4.06 5.89 5.05
2	Ppk analysis in Minitab • Run a capability analysis in Minitab: Stat/ quality tools/ capability Sixpack/Normal (see the prior job aid for detailed steps). • The LSL is 5, and the USL is 30.	Capability Sixpack (Normal Distribution) ✕ Data are arranged as ⦿ Single column: [tensile strength' Subgroup size: [1 (use a constant or an ID column) ○ Subgroups across rows of: Lower spec: [5 Upper spec: [30 Historical mean: [ ] (optional) Historical standard deviation: [ ] (optional) Transform...　Tests...　Estimate...　Options... Select　Help　OK　Cancel

Step	Instructions	Illustration
3	Interpreting the initial results • Note that the data is not normal. • Also, process is not stable. • Therefore, the Ppk calculation is not considered reliable (although it is well less than 1 and the histogram shows the data is outside the USL and LSL, so it is reasonable to assume already the process is not capable). • Try a data transformation.	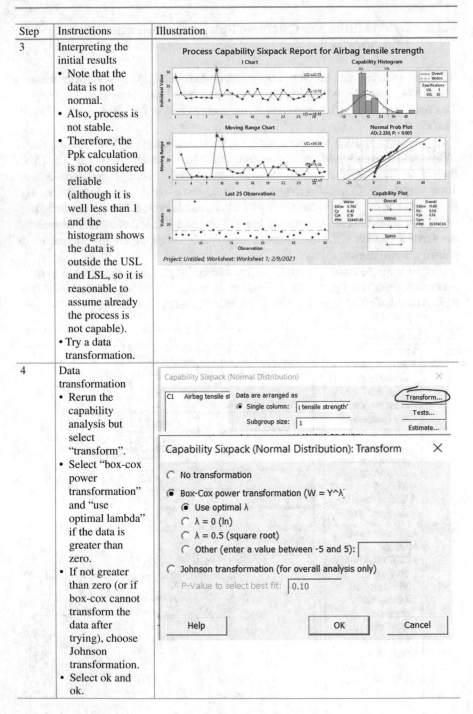
4	Data transformation • Rerun the capability analysis but select "transform". • Select "box-cox power transformation" and "use optimal lambda" if the data is greater than zero. • If not greater than zero (or if box-cox cannot transform the data after trying), choose Johnson transformation. • Select ok and ok.	

Step	Instructions	Illustration
5	Transformation outcomes • Note the assumptions for a stable process and normal data are now met. • The Ppk value is 0.08 which is less than 1 and the preferred 1.33, so the process is not considered capable.	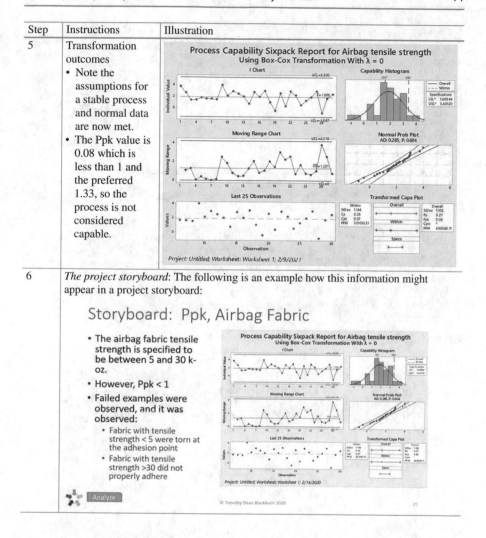
6	*The project storyboard*: The following is an example how this information might appear in a project storyboard:	

4.3.4 Discrete Data Process Capability

Remember earlier in the book we observed that KIND Karz is seeing 370 defects per 1000 airbags produced after the new assembler equipment was installed (see Fig. 4.25). This is an example of discrete data. What is the Ppk? The normal six pack may not work well in this case. Later, a job aid will provide step-by-step instructions on how to calculate Ppk when this occurs.

But first, let's consider a bit of theory behind the approach. One way to estimate Ppk in this case is the DPMO method. DPMO is defects per million opportunities. It is called the DPMO method as reference tables are available that associate DPMO with Ppk. While this book won't use DPMO directly to estimate Ppk, it uses the same underlying approach.

In general, this method assumes the defect proportions are normally distributed. If so, then the Z value can be determined using a standard normal curve. (If you

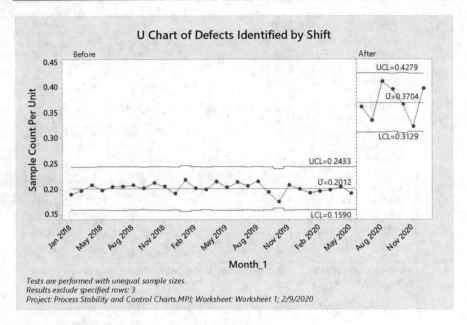

Fig. 4.25 Airbag warranty claims and recalls control chart

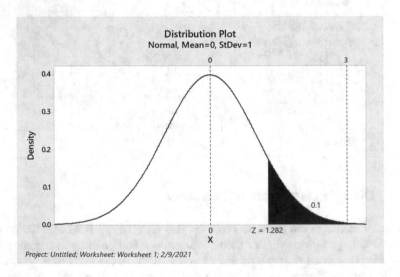

Fig. 4.26 Example defect rate Z value

know the defect data is not normally distributed, other distributions can be used with a similar approach.) For example, consider a situation where 100 items are produced and 90 are acceptable. That is, 10% are defective. In Minitab, we can easily determine then the Z value as shown in Fig. 4.26. Note the Z value is 1.282. Z is the number of standard deviations from the mean.

Since the Z is within three sigma, we would expect the Ppk to be less than 1 (the Z would need to be 3 for Ppk = 1). So, to determine Ppk, simply divide the calculated Z by 3, or in this case 1.282/3 = 0.4273 (Bothe, 1997, pp. 173, 533, 537).

Also Process Sigma can now be easily calculated, which is Z + 1.5 to account for an expected shift of the process sigma over time (George et al., 2005). In this case, 1.282 + 1.5 = 2.782. This and the above are summarized in Table 4.11.

In the above example, we assumed a binomial or pass/fail event, or each item was just listed as pass/fail. In some cases, it might resemble more of a Poisson event, or each item could have multiple defect types. For example, you might be inspecting bottles on a line. You might be checking for proper label application, cracks, and bottle cap. In this case, there are three defect types per bottle. These are described as defect opportunities per unit. When this occurs, the proportion used to calculate Z, or defects per opportunity (DPO), is calculated as the number of defects divided by the number of units times the defect opportunity per unit. This will be demonstrated further in the job aid.

Table 4.11   Example defect rate example table

Category	Entry or Results
Number of units	100
Defects	10
Defects opportunity per unit	1
Defects per opportunity	0.100
Process sigma	2.78
PpK	**0.43**

4.3.5   Calculating Ppk for Discrete Data in Minitab

Step	Instructions	Illustration
1	Calculate DPO • From the KIND Karz example, recall three are 370 warranty claims or recalls per 1000 cars produced. • There are 5 defect opportunities per airbag (e.g., pinholes, adhesion, smudging, meeting thickness limits, meeting dimensions). • Calculate the DPO.	• Number of units: 1000. • Defects: 370. • Defect opportunity per unit: 5. • Defects per opportunity (DPO): (370)/(1000*5) = 0.074 Note: It is recommended to have at least 100 samples and at least 5 defects to use this method

Step	Instructions	Illustration
2	Calculate the Z value • In Minitab: Graph/ probability distribution plot/ view probability/ shaded area/right tail. • Select "probability" and enter the DPO in the box (0.074). Select "ok". • Note the Z value is 1.447. • Calculate Ppk and process sigma (optional) as shown. Note: Excel can also calculate the Z value for you by using the following: NORMSINV(1-DPO)	**Distribution Plot** Normal, Mean=0, StDev=1 _Project: Untitled; Worksheet: Worksheet 1; 2/9/2021_ • Ppk = Z/3 = 1.447/3 = 0.48. • Process sigma = Z + 1.5 = 1.447 + 1.5 = 2.95.
3	Summarizing the results	

Category	Entry or Results
Number of Units	1,000
Defects	370
Defect opportunity per Unit	5
Defects per Opportunity	0.074
Process Sigma	2.95
Ppk	**0.48**

4   _The project storyboard_: The following is an example how this information might appear in a project storyboard:

Storyboard: Ppk for Airbag defect rate

• Top: Currently not capable (Ppk = 0.48)

• Objective: To be below 100 defects found per 1,000 airbags.

• Below: Ppk goal improve to at least 0.68 for defect rate

Category	Entry or Results
Number of Units	1,000
Defects	370
Defect opportunity per Unit	5
Defects per Opportunity	0.074
Process Sigma	2.95
Ppk	0.48

Category	Entry or Results
Number of Units	1,000
Defects	100
Defect opportunity per Unit	5
Defects per Opportunity	0.020
Process Sigma	3.55
Ppk	0.68

 Measure

© Timothy Dean Blackburn 2020   25

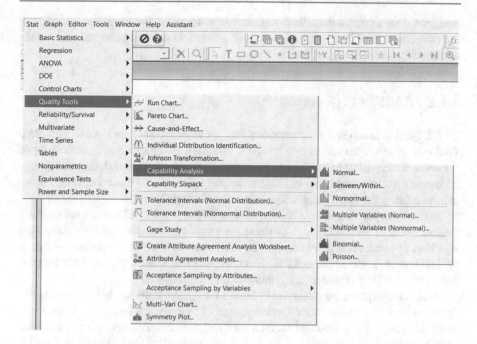

Fig. 4.27 Minitab binomial and Poisson process capability tools

Another alternative is to use the Binomial or Poisson process capability tools in Minitab. These are located under Stat/Quality Tools/Capability analysis as shown in Fig. 4.27. Use binomial when there is one defect opportunity per unit and Poisson when there is more than one defect opportunity per unit. While these analysis methods are beyond the scope of the book, they can provide valuable insights as to the capability of the process

4.4 Measurement System Analysis (with Gage R&R, Attribute Agreement Analysis): Minitab Methods and Analysis Detail

4.4.1 Introduction to Gage R&R and Attribute Agreement Analysis

In a production process, the variation comes from the process itself but also from the measurement system. If we don't have a reliable and accurate measurement system, we can't be confident in the process stability or capability analysis outcomes. In practical terms, we don't really know what our process is doing if there is too much variation in the measurement system.

This section will provide step-by-step instructions in Minitab how to run and interpret a GAGE R&R and Attribute Agreement Analysis using the KIND Karz case study. A GAGE R&R is for continuous data, and an Attribute Agreement

Analysis is for attribute data. Both will enable us to assess the degree of variation in the data that comes from our measurement system, between operators, and within operators (when they repeat the same test).

4.4.2   GAGE R&R: Overview

GAGE R&R is to be used for continuous data, or when data can be logically divided. This can include measurements, weights, volumes, or other non-discrete units of measure. GAGE R&R identifies issues with repeatability (the first R, or same operator repeating a measurement) and reproducibility (the second R, or a different operator measuring the same thing when it is a non-destructive test).

To illustrate practically how to apply GAGE R&R, the KIND Karz, Brake caliper torsion test will be used, which is measuring torsion, a continuous variable. This will help us understand if we can rely on the Brake caliper test. Important as this is the last line of defense before KIND Karz receives the Brake calipers from the vendors. But first, let's consider key principles and approaches.

First, the design of the analysis is paramount if the results are to be meaningful. Select two or three assessors (operators) and choose five to ten items to be measured. Ensure they are marked in such a way you can identify them, but it should not be apparent to the operators (e.g., long random number that look similar, or facilitator should hand them out one at a time). Design the order such that operators measure each item two (or three) times in random sequence. Be there to observe (not as a participant but to identify issues that might arise that could affect or bias outcomes).

Another important criterion to determine is whether it is a crossed or nested study. For crossed, the item being measured is not destroyed, so assessors can measure it multiple times and across different assessors. With nested, assessors measure unique parts (can also occur during destructive testing or test to failure). See Fig. 4.28 for an illustration of crossed versus nested designs. The example we will use for KIND Karz will be assumed to be nested—the calipers won't be tested to failure.

For the KIND Karz example, recall that brake calipers have been determined to be a major source of warrantee claims recently. Failures are occurring, but tests for the same calipers passed at the vendor. Can we trust the measurement system which provides early indications whether a caliper could become or is defective?

The design for this GAGE study will require ten parts, selected to represent the expected range of the process variation. Then three assessors will measure the ten parts, tested three times per part, in a random order. A crossed gage R&R study will be performed to assess the variability in measurements that might be from the measurement system. The torsion resistance must be at least 80 N-m for the test to pass. In addition to other key criteria, we hope to see that most variation is in the parts and the percent of Gage variation should be low.

- Less than 10%: Good – indicates little variation in your data is due to the measurement system and most variation is true.
- 10–30%: Don't accept anything more than 30.
- More than 30%: Unacceptable – the measurement system is too unpredictable. Reproducibility and repeatability should also be less than <30%.

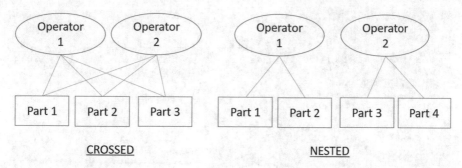

Fig. 4.28 Crossed versus nested designs

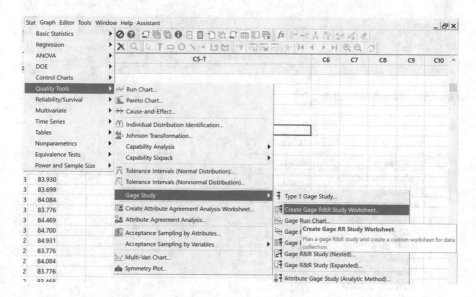

Fig. 4.29 Designing a GAGE study

4.4.3   Designing and Analyzing the GAGE R&R Study in Minitab

To design the study, the Minitab path is Stat/Quality Tools/Gage Study/Create Gage R&R Study Worksheet (Minitab Online Help https://www.minitab.com/en-us/support/). See Fig. 4.29 for a graphic of the steps. Following the graphic, step-by-step instructions are provided.

Once the data is gathered, the analysis can be completed by Stat/Quality Tools/ Gage Study/Gage R&R Study (crossed). The path for nested is similar but out of scope for this book. See Fig. 4.30 for an illustration of the path. The analysis steps are also included in the job aid that follows.

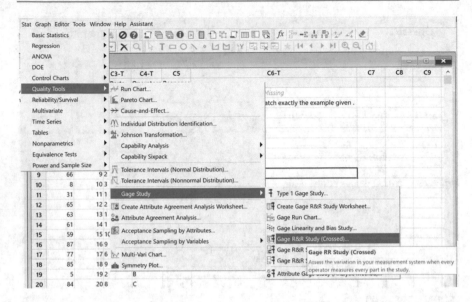

Fig. 4.30 Analyzing the study

Step	Instructions	Illustration
1	Defining the GAGE R&R design • Minitab path: Stat/ quality tools/gage study/create Gage R&R Study Worksheet. • Enter "10" for "number of parts". • Enter "3" for "number of replicates." this is the number of times an operator will test each part. • Enter "3" for "number of operators". • Enter operator aliases (or names, although this can be sensitive) under "operator names". • Select "options".	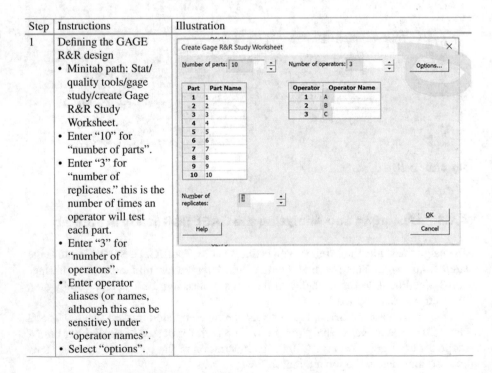

Step	Instructions	Illustration
2	• Check the box for "randomize all runs". • Check the box for "store standard run order in worksheet".	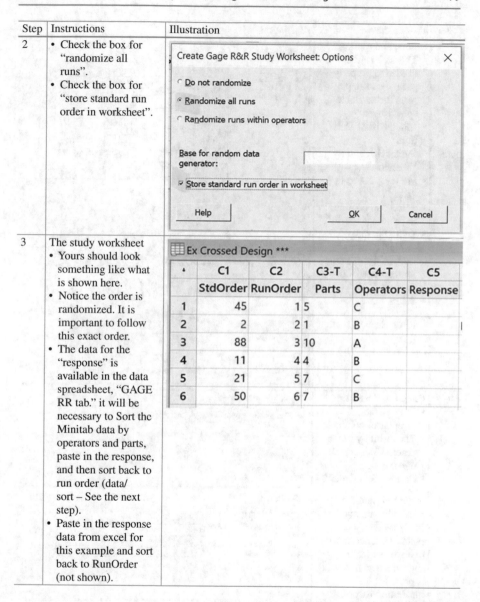
3	The study worksheet • Yours should look something like what is shown here. • Notice the order is randomized. It is important to follow this exact order. • The data for the "response" is available in the data spreadsheet, "GAGE RR tab." it will be necessary to Sort the Minitab data by operators and parts, paste in the response, and then sort back to run order (data/ sort – See the next step). • Paste in the response data from excel for this example and sort back to RunOrder (not shown).	

Ex Crossed Design ***

	C1 StdOrder	C2 RunOrder	C3-T Parts	C4-T Operators	C5 Response
1	45	1	5	C	
2	2	2	1	B	
3	88	3	10	A	
4	11	4	4	B	
5	21	5	7	C	
6	50	6	7	B	

Step	Instructions	Illustration
4	Sorting and pasting • This step is just needed to properly paste the example from the available spreadsheet. • Minitab path: Data/ Sort. • For level 1, sort by "operators". • For level 2, sort by "parts". • Select "all columns" in "columns to sort". • Select "in the original columns" for "storage location for the sorted columns". • Select "ok".	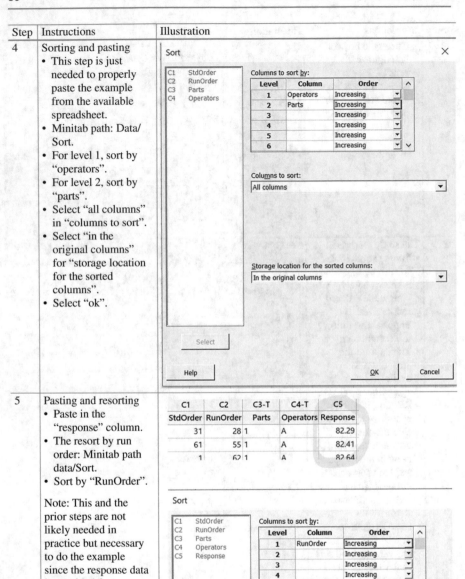
5	Pasting and resorting • Paste in the "response" column. • The resort by run order: Minitab path data/Sort. • Sort by "RunOrder". Note: This and the prior steps are not likely needed in practice but necessary to do the example since the response data is provided for illustration purposes	

Step	Instructions	Illustration
6	Analysis • Minitab path: Stat/ quality tools/gage study/Gage R&R Study (crossed). • For "part numbers" identify the item being measured. In this case, the column heading is also "parts". • Enter the names of the operators in "operator" (in this case the column heading is also entitled "operators"). • In "measurement data" select "response". • Select "gage info" if you would like to add further description of the study. • Select "options" if you would like to also check for tolerance (not included in this example but shown in the next step in case you need this in practice).	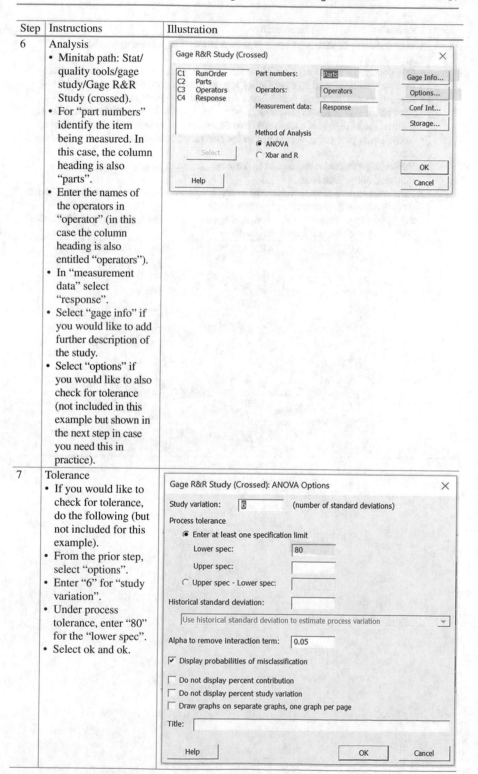
7	Tolerance • If you would like to check for tolerance, do the following (but not included for this example). • From the prior step, select "options". • Enter "6" for "study variation". • Under process tolerance, enter "80" for the "lower spec". • Select ok and ok.	

Step	Instructions	Illustration
8	The components of variations graph is shown below. The following are the interpretations: • Components of variation graph: Most variation is part to part (good), while overall gage, repeatability, and reproducibility appear to be less than 30—Good. • R chart: Would like to see points in the control limits—Most are. • Xbar chart: Would like to see majority of points (parts values) outside control limits (narrow limits represent a more effective measurement system)—True in this case. • Response by part graph: Would like to see full range represented. Assumed to be true here. • Response by assessor: Would like to see near horizontal alignment suggesting reproducibility. True here. • Part*assessor interaction: Would like to see lines parallel, which is true here. Indicates there is no significant part/people interaction. • So, by just looking at the graph, it would appear the testing approach is good. But let's confirm by looking at the analysis results.	

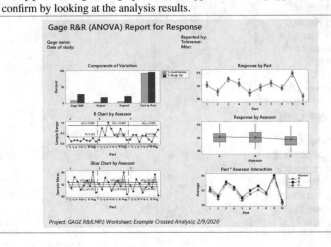

Step	Instructions	Illustration
9	Analysis results – Significance • The part P-value should be less than 0.05, indicating the measurement system is detecting a difference between parts (shown as part (operator) in a nested study). • The operator P-value should be greater than 0.05, indicating the there is no significant operator effect. Not true here. • The part*operator P-value should be *greater* than 0.05, indicating there isn't a significant interaction between parts and operators (N/A for nested as measurements are not repeated for the same part). • In this example, there is an operator effect.	**Two-Way ANOVA Table With Interaction** | Source	DF	SS	MS	F	P
--------	----	----	----	----	----
Parts	9	88.3619	9.81799	492.291	0.000
Operators	2	3.1673	1.58363	79.406	0.000
Parts * Operators	18	0.3590	0.01994	0.434	0.974
Repeatability	60	2.7589	0.04598		
Total	89	94.6471				 *α to remove interaction term = 0.05*

Step	Instructions	Illustration
10	Analysis – GAGE R&R outcomes • For the variance components, largest percentage should be parts versus gage elements. • For gage evaluation. ◦ % contribution should be less than 9%. We have 7.76%, which is good ◦ % study variation (and % tolerance if used) should be less than 30% to indicate gage is reliable. That is true here (27.86%)	**Gage R&R** **Variance Components**

Gage R&R

Variance Components

Source	VarComp	%Contribution (of VarComp)
Total Gage R&R	0.09143	7.76
Repeatability	0.03997	3.39
Reproducibility	0.05146	4.37
Assessor	0.05146	4.37
Part-To-Part	1.08645	92.24
Total Variation	1.17788	100.00

Gage Evaluation

Source	StdDev (SD)	Study Var (6 × SD)	%Study Var (%SV)
Total Gage R&R	0.30237	1.81423	27.86
Repeatability	0.19993	1.19960	18.42
Reproducibility	0.22684	1.36103	20.90
Assessor	0.22684	1.36103	20.90
Part-To-Part	1.04233	6.25396	96.04
Total Variation	1.08530	6.51180	100.00

Number of Distinct Categories = 4

Number of distinct categories should be 5 or more
• In this case, the number of categories the test can differentiate is of concern.

Step	Instructions	Illustration
11	Run chart • Minitab path: Stat/ quality tools/gage study/gage run chart. • Enter "parts" in "pater numbers". • Enter "operators" in "operators". • Enter "response" in "measurement data".	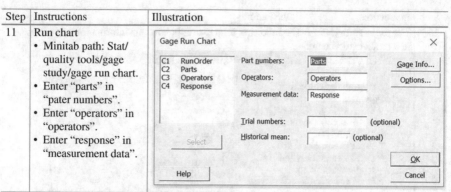

12	Run chart interpretation
	 • Ideal: Low variation between people and trials.
	 • Ideal: Variation detected between parts.
	 • Note: Each dot represents a single measurement by the same person and is connected by a line.
	   ◦ Connected dots should not vary (repeatability).
	   ◦ Comparing dots for a given part should not vary (reproducibility).
	 • In this case, the run chart gives favorable results visually.
	 • Not applicable for nested—Cannot measure the same part multiple times.

Gage Run Chart of Response by Parts, Operators

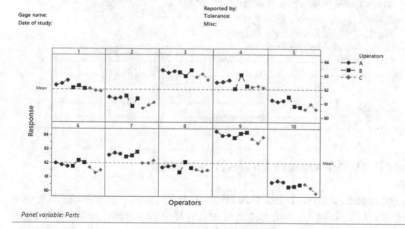

Step	Instructions	Illustration
13	*The project storyboard*: The following is an example how this information might appear in a project storyboard: 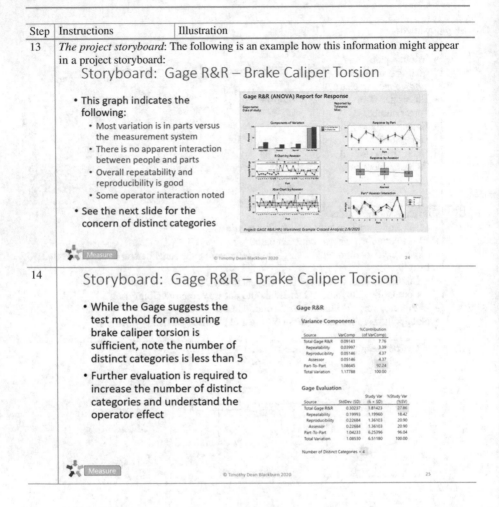	

4.4.4   Attribute Agreement Analysis in Minitab

Attribute Agreement Analysis is like GAGE R&R but is used for attribute data versus continuous data. Like GAGE, it can confirm if your measurement system is reliable and will identify issues with repeatability (same operator repeating a measurement) and reproducibility (different operator measuring the same thing).

An Attribute Agreement Analysis is performed when attributed (or ordinal) measurements rely on subjective classifications (or ratings). This can include such subjective measurements as good/bad, go/no go, pass/fail, as well as multiple characteristics. It can sometimes also apply to continuous data when there is a pass/fail decision (e.g., when greater than a certain thickness).

There are some general guidelines needed to design the study and perform the analysis, including good practices. Individuals assigned as raters will select passing and not passing examples or identify multiple failing attributes. There should be at least two raters, performing blind studies (they should be unaware of the specific samples and identified defects, or which they are reinspecting). Each rater should evaluate the samples at least twice but should be unaware they are doing so. Find a way to identify samples without making the rater aware (e.g., long random identifiers containing numbers, letters, and symbols). Also randomize runs to further reduce bias.

If there are only two categories (good/bad, pass/fail, etc.), have a minimum of 20 good and 20 bad (or failing) for comparison (40 total). However, Minitab recommends at least 50 samples generally. If there are more than two categories, have equal quantities of passing/not-passing ideally for each sample.

The primary criterion to determine the measurement system effectiveness is the Kappa value. It is calculated as "the ratio of the proportion of times that the appraisers agree (corrected for chance agreement) to the maximum proportion of times that the appraisers could agree (corrected for chance agreement)" (Minitab Online Help https://www.minitab.com/en-us/support/). A Kappa value of 0.7 is considered the minimum (the Kappa value should be at or above 0.7 to conclude the measurement system is adequate). A Kappa value of 0.9 might be needed for more critical inspection activities.

The equation is as follows:

$$K = \left(P_{observed} - P_{chance}\right) / \left(1 - P_{chance}\right)$$

where
$P_{observed}$ = proportions of units on which both raters agree.
P_{chance} = proportion of agreements expected by chance.

In this section, we will use another example from KIND Karz, specifically related to airbag defect identification. Airbags have been determined to be a major source of recalls at KIND Karz. Failures are occurring, but visual inspections for the same bags passed at the vendor.

Can we trust the measurement system which provides early indications whether an airbag could become or is defective?

First let's define what attributes the inspectors are considering:
- Web thickness: It must exceed a standard caliper setting of 1 mm (actual measurement is not indicated in the data but rather only pass or fail if it is over or under the pre-set value).
- Discoloration based on photo images.
- Pinholes – are there any pinholes using the pressure test device?
- Missing adhesive – by observation, is there any missing adhesive at the primary location?
- Overlap – overlap of the fabric at the primary connection point must meet or exceed the scale template.

Inspectors log how many of the of the above passed. All five tests must pass for the airbag to be accepted. Assessors note how many tests each airbag passes. Note: we could have just had a pass/fail—that is, if any of the above failed, the test would have failed overall. But this approach gives greater resolution and data for further evaluation and a broader scope for purposes of illustration.

Next comes the challenge of choosing the assessors (or inspectors in this case). There should be at least two as noted above. Ideally, the inspectors should have a range of experience and represent the various work areas (if different). For the KIND Karz example, four operators were chosen (John, Sheila, Timothy, Leslie).

Now it's time to design the study. While Minitab has a feature to design the worksheet, the parameters need to be determined first. For multiple attributes (up to five), Albert chose the standard guidance of 50 samples, reasonably distributed across defect levels. Based on experience, samples were created or selected with known defects. Usually samples are balanced, but not necessarily if there is a reason to do otherwise (the analysis will still run). In this case, Albert apportioned based on the known history of distribution of passes (one to four attributes passing) as shown in Table 4.12.

The Minitab path to design the study is Stat/Quality Tools/Create Attribute Analysis Worksheet. This is further illustrated in Fig. 4.31. The path to analyze the study is Stat/Quality tools/Attribute Agreement Analysis and is illustrated in Fig. 4.32. The job aid to create the study worksheet and perform the analysis is shown below the figures.

Table 4.12 Sample selection

Attributes passing	Quantity
1	10
2	12
3	10
4	8
5	10
Total	50

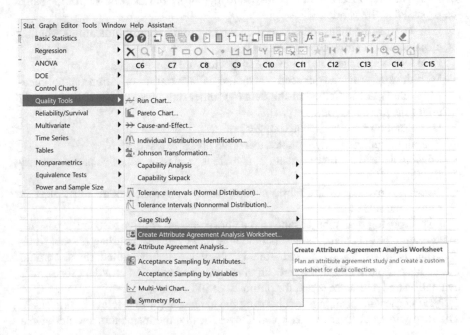

Fig. 4.31 Minitab path, attribute study worksheet

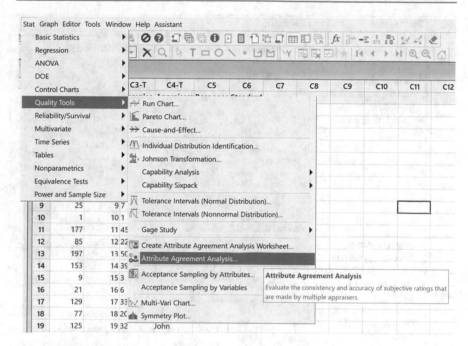

Fig. 4.32 Minitab path to analyze an attribute agreement analysis

Step	Instructions	Illustration
1	Designing the study • Stat/quality tools/ create attribute analysis worksheet. • Choose "sample standard/attribute in numeric values". • Enter 50 for "number of samples". • Enter 4 for "number of appraisers". • Enter names (or aliases) in "appraiser names". • Select 2 for "number of replicates. • Click "options". Note: These will vary depending on the design parameters for a particular situation	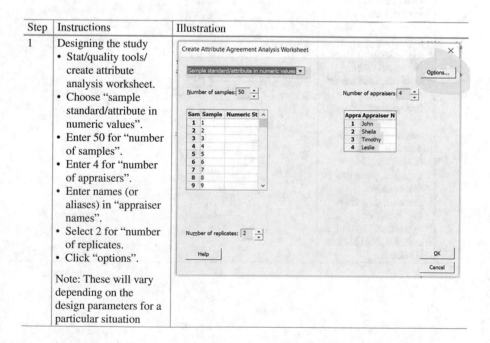

Step	Instructions	Illustration
2	Other parameters • Select "randomize all runs". • Select "store standard run order in worksheet". • Select ok and ok.	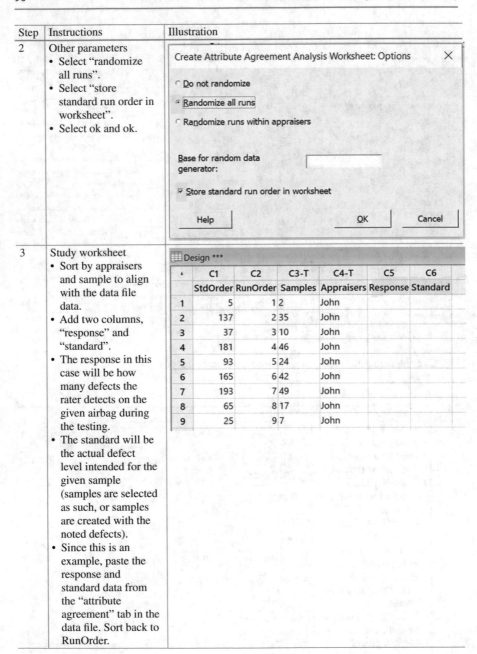
3	Study worksheet • Sort by appraisers and sample to align with the data file data. • Add two columns, "response" and "standard". • The response in this case will be how many defects the rater detects on the given airbag during the testing. • The standard will be the actual defect level intended for the given sample (samples are selected as such, or samples are created with the noted defects). • Since this is an example, paste the response and standard data from the "attribute agreement" tab in the data file. Sort back to RunOrder.	

For step 3, the illustration shows:

Design ***

•	C1 StdOrder	C2 RunOrder	C3-T Samples	C4-T Appraisers	C5 Response	C6 Standard
1	5	1	2	John		
2	137	2	35	John		
3	37	3	10	John		
4	181	4	46	John		
5	93	5	24	John		
6	165	6	42	John		
7	193	7	49	John		
8	65	8	17	John		
9	25	9	7	John		

Step	Instructions	Illustration
4	Analysis • Stat/quality tools/ attribute agreement analysis. • Select "response" for "attribute column". • Select "sample" for "samples". • Select "appraisers" for "appraisers". • Select "standard" for the "known standard/ attribute". • Select ok.	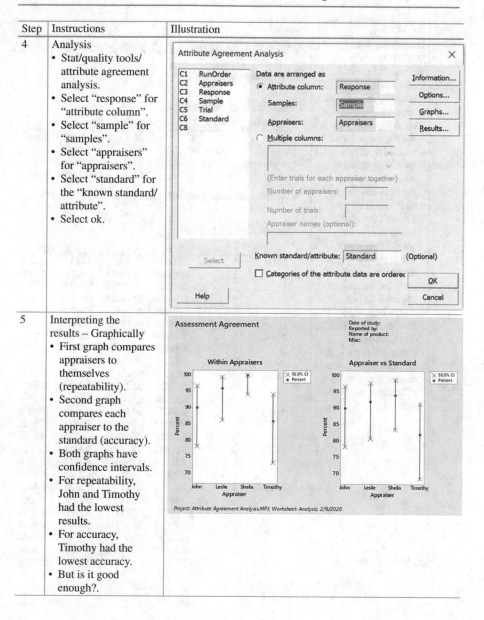
5	Interpreting the results – Graphically • First graph compares appraisers to themselves (repeatability). • Second graph compares each appraiser to the standard (accuracy). • Both graphs have confidence intervals. • For repeatability, John and Timothy had the lowest results. • For accuracy, Timothy had the lowest accuracy. • But is it good enough?.	

Step	Instructions	Illustration
6	Interpreting the Minitab output – Within appraisers • The Fleiss kappa statistics here shows repeatability, or accuracy when an operator repeats the visual inspection (blind, or without their knowledge it is the same item). • In this case all Kappa's are at or above 0.7, so we would assume the test is repeatable, but for more critical systems, one might expect a higher kappa (e.g., 0.9). • However, Timothy has the lowest percent accuracy, as reflected also in the lowest kappa values (although still at or above 0.7).	See illustration below

Attribute Agreement Analysis for Response

Within Appraisers

Assessment Agreement

Appraiser	# Inspected	# Matched	Percent	95% CI
John	50	45	90.00	(78.19, 96.67)
Leslie	50	48	96.00	(86.29, 99.51)
Sheila	50	50	100.00	(94.18, 100.00)
Timothy	50	43	86.00	(73.26, 94.18)

\# Matched: Appraiser agrees with him/herself across trials.

Fleiss' Kappa Statistics

Appraiser	Response	Kappa	SE Kappa	Z	P(vs > 0)
John	1	1.00000	0.141421	7.0711	0.0000
	2	0.83060	0.141421	5.8733	0.0000
	3	0.81917	0.141421	5.7924	0.0000
	4	0.86450	0.141421	6.1129	0.0000
	5	0.86450	0.141421	6.1129	0.0000
	Overall	0.87472	0.070945	12.3295	0.0000
Leslie	1	1.00000	0.141421	7.0711	0.0000
	2	0.89605	0.141421	6.3360	0.0000
	3	0.86450	0.141421	6.1129	0.0000
	4	1.00000	0.141421	7.0711	0.0000
	5	1.00000	0.141421	7.0711	0.0000
	Overall	0.94965	0.071401	13.3002	0.0000
Sheila	1	1.00000	0.141421	7.0711	0.0000
	2	1.00000	0.141421	7.0711	0.0000
	3	1.00000	0.141421	7.0711	0.0000
	4	1.00000	0.141421	7.0711	0.0000
	5	1.00000	0.141421	7.0711	0.0000
	Overall	1.00000	0.071052	14.0741	0.0000
Timothy	1	0.83060	0.141421	5.8733	0.0000
	2	0.84000	0.141421	5.9397	0.0000
	3	0.70238	0.141421	4.9666	0.0000
	4	0.70238	0.141421	4.9666	0.0000
	5	1.00000	0.141421	7.0711	0.0000
	Overall	0.82354	0.071591	11.5034	0.0000

Step	Instructions	Illustration
7	Interpreting the Minitab output – Appraisers versus the standard • This section, "each appraiser vs. standard," assesses the accuracy. • Notice all kappa values are above 0.7. • So, the test is considered sufficient for accurately assessing the types of defects at the standard cutoff of 0.7 (but not at 0.9).	(see below)

Each Appraiser vs Standard

Assessment Agreement

Appraiser	# Inspected	# Matched	Percent	95% CI
John	50	45	90.00	(78.19, 96.67)
Leslie	50	46	92.00	(80.77, 97.78)
Sheila	50	47	94.00	(83.45, 98.75)
Timothy	50	41	82.00	(68.56, 91.42)

Matched: Appraiser's assessment across trials agrees with the known standard.

Fleiss' Kappa Statistics

Appraiser	Response	Kappa	SE Kappa	Z	P(vs > 0)
John	1	1.00000	0.100000	10.0000	0.0000
	2	0.91694	0.100000	9.1694	0.0000
	3	0.90736	0.100000	9.0736	0.0000
	4	0.92913	0.100000	9.2913	0.0000
	5	0.93502	0.100000	9.3502	0.0000
	Overall	0.93732	0.050211	18.6674	0.0000
Leslie	1	1.00000	0.100000	10.0000	0.0000
	2	0.83838	0.100000	8.3838	0.0000
	3	0.80725	0.100000	8.0725	0.0000
	4	1.00000	0.100000	10.0000	0.0000
	5	1.00000	0.100000	10.0000	0.0000
	Overall	0.92462	0.050396	18.3473	0.0000
Sheila	1	1.00000	0.100000	10.0000	0.0000
	2	0.83060	0.100000	8.3060	0.0000
	3	0.81917	0.100000	8.1917	0.0000
	4	1.00000	0.100000	10.0000	0.0000
	5	1.00000	0.100000	10.0000	0.0000
	Overall	0.92476	0.050257	18.4006	0.0000
Timothy	1	0.91159	0.100000	9.1159	0.0000
	2	0.81035	0.100000	8.1035	0.0000
	3	0.72619	0.100000	7.2619	0.0000
	4	0.84919	0.100000	8.4919	0.0000
	5	1.00000	0.100000	10.0000	0.0000
	Overall	0.86163	0.050500	17.0622	0.0000

Step	Instructions	Illustration
8	Interpreting the Minitab output – Between appraisers • This section assesses reproducibility, or whether different appraisers get the same results on the same sample. • Here all the kappa values are greater than 0.7 (but some less than 0.9). • So, the test is considered sufficient for reproducibility at standard kappa 0.7.	(see below)

Between Appraisers

Assessment Agreement

# Inspected	# Matched	Percent	95% CI
50	37	74.00	(59.66, 85.37)

Matched: All appraisers' assessments agree with each other.

Fleiss' Kappa Statistics

Response	Kappa	SE Kappa	Z	P(vs > 0)
1	0.954392	0.0267261	35.7101	0.0000
2	0.827694	0.0267261	30.9695	0.0000
3	0.772541	0.0267261	28.9058	0.0000
4	0.891127	0.0267261	33.3429	0.0000
5	0.968148	0.0267261	36.2248	0.0000
Overall	0.881705	0.0134362	65.6218	0.0000

Step	Instructions	Illustration
9	Interpreting the Minitab output – Overall attribute agreement analysis • This section summarizes the overall attribute agreement analysis. • Note that all Kappa's are >0.70. • However, it is 74% accurate, less than we would like for a life critical system. • And the 95% confidence interval for accuracy goes as low as 59.66% (likely more a result of the sample size—a tighter CI would require more samples).	**All Appraisers vs Standard** **Assessment Agreement** 	# Inspected	# Matched	Percent	95% CI
---	---	---	---
50	37	74.00	(59.66, 85.37) | *# Matched: All appraisers' assessments agree with the known standard.* **Fleiss' Kappa Statistics** | Response	Kappa	SE Kappa	Z	P(vs > 0)
---	---	---	---	---
1	0.977897	0.0500000	19.5579	0.0000
2	0.849068	0.0500000	16.9814	0.0000
3	0.814992	0.0500000	16.2998	0.0000
4	0.944580	0.0500000	18.8916	0.0000
5	0.983756	0.0500000	19.6751	0.0000
Overall	0.912082	0.0251705	36.2362	0.0000	
10	Practical considerations • In this case, the results showed the test method is adequate for a standard kappa at 0.7. • However, the percent accuracy (and CI) is lower than we would prefer for a life-critical system. • Further review by the team revealed the pinhole test was the least accurate. • Given some bags passed the test (and were installed in cars but failed later in an accident), a new test was designed and shown to be 98% reliable to predict pinhole failures.	
11	*The project storyboard*: The following is an example how this information might appear in a project storyboard: ## Attribute Agreement Analysis: Airbags • Attribute Agreement Analysis was sufficient for test method • However, overall accuracy less than desired • Further review indicated the highest incidence of misses was due to the pin hole test • A new test method will need to be designed for the pin hole test Measure © Timothy Dean Blackburn 2020	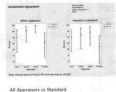 All Appraisers vs Standard Assessment Agreement Fleiss' Kappa Statistics 31

4.5   Pareto Analysis: Minitab Methods and Analysis Detail

4.5.1   Creating a Pareto Chart

A Pareto chart is a type of bar chart, where the X axis represents categories and the bars represent counts. The bars are arranged from highest to smallest counts. This allows us to look for situations where ~20% of the categories represent ~80% of the problem, and focus on what is important, or isolate areas where root causes might exist.

Vilfredo Pareto (an economist-sociologist) noticed 20% of the people owned 80% of the wealth, which was published in the book *Cours d'Économie Politique* (1896/97). Later, George Zipf and Joseph Juran further refined, developed, and made Pareto's theories applicable to business and management (mid-twentieth century)   (https://www.businessballs.com/self-management/paretos-80-20-rule-theory/).

Before we describe how to build a Pareto chart, let's look and differentiate two examples. Imagine you are managing a Six Sigma project and your objective is to reduce defects in your small engine and drive train department. You create a Pareto chart based on defects by component, as shown in Fig. 4.33.

What would you conclude from this? One possibility could be there really is no difference in the distribution of problems. However, look for another way to classify your data.

Fortunately, your defect data is also indicated by machine center (see Fig. 4.34). Looking at the same data, but stratified in a different way, what would your conclusion now be? What would be your next step? Obviously, you would want to see what is different at the casting station that is causing an apparent significantly higher defect count.

Reading the Pareto chart is quite easy and immediately provides visual insights. The line with dots at the top is the cumulative number of counts, up to 100% as shown on the right Y axis. The left Y axis is the count, and the X axis categorizes by stratification factors. Under the chart, Minitab shows the count for the given category, the category percent of total, and the cumulative percentage. The count of cumulative percentage does not have to be exactly 80% to represent a Pareto effect, however. Just look for one to three bars that are visually higher than the remainder.

To illustrate how to construct a Pareto in Minitab, let's return to the KIND Karz case study. They are having an unexpected increase in recalls and warrantee claims. Fortunately, Albert has data from the last quarter across a variety of categories. These include airbags, brakes, defect, electronics, emissions, hardware, and propulsion. From this, he will be able to construct a Pareto chart to see if one or more categories indicates a Pareto effect, which will allow an opportunity to focus on those categories first. The path to run a Pareto chart in Minitab is Stat/Quality Tools/Pareto Chart as shown is Fig. 4.35.

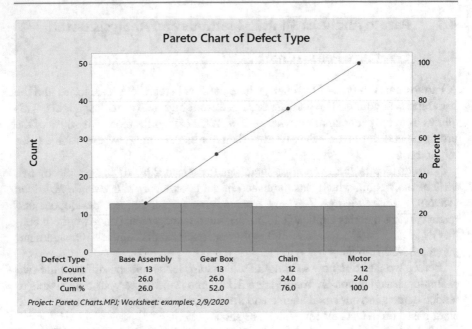

Fig. 4.33   Pareto chart – no apparent Pareto effect

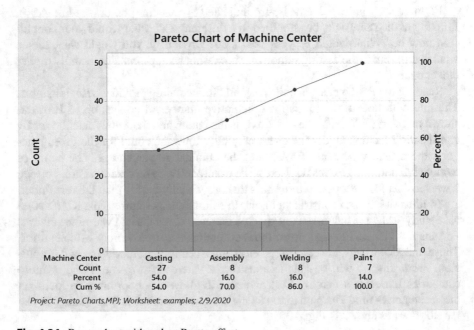

Fig. 4.34   Pareto chart with a clear Pareto effect

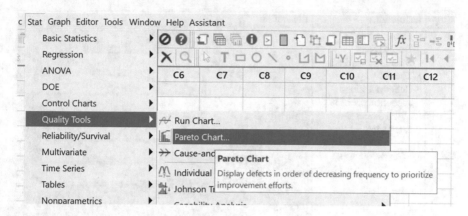

Fig. 4.35 Minitab path, Pareto

4.5.2   Constructing a Pareto Chart in Minitab

Step	Instructions	Illustration
1	The data • Paste the data into Minitab from the "Pareto" tab in the data spreadsheet.	
2	Creating the Pareto chart • Stat/quality tools/ Pareto chart. • Select "defect" in "defects or attribute data in". • Select "do not combine." if you do not select this, sometimes it will put small categories in an "other" bar. • Select "OK".	

Step	Instructions	Illustration
3	• There is a clear Pareto effect—Brakes and airbags. • These two categories alone account for 79.1% of the defect categories.	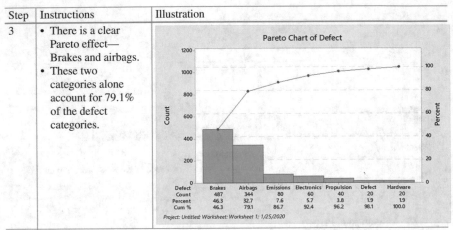
4	*The project storyboard*: The following is an example how this information might appear in a project storyboard: ### Pareto Analysis by Defect Results 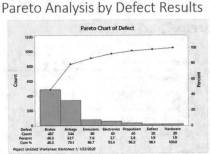 • Brakes and Airbags account for 79.1 percent of all defects • This will carry forward to Analyze to get to the root cause, and will be the primary focus areas 

4.6   Test of Proportions: Minitab Methods and Analysis Detail

4.6.1   Introduction to Test of Proportions

Often, it is necessary to compare two or more samples of proportional data. These can include such things as percent defective, yield, or other measures that include a numerator and denominator. This can occur at various DMAIC phases. For example, in the Measure and Analyze phases, we might need to determine if a category is worst (or better) than another. Then, in the Improve or Control phase, we might want to compare before and after and determine if an improvement occurred.

There are two methods presented in this section—the first is the test of two proportions and the other Chi-Square test of multiple proportions. Both have a similar null and alternative hypothesis. The null hypothesis (H_o) states that the proportions are the same. The Alternative (H_A) states that at least one proportion is different. If $P \geq 0.05$, do not reject H_o. That is, there is insufficient statistical evidence to conclude one proportion is different from the another. (However, one cannot claim they are equal in this case.) But if $P < 0.05$, reject H_o and accept the H_A that at least one proportion is different from another.

To describe how to perform the analysis in Minitab, we will consider two examples from the KIND Karz case study.

4.6.2   Test of Two Proportions in Minitab

KIND Karz has been experiencing issues with airbags and is seeing an increase in warrantee work. Is one vendor worse than another? Can you tell by looking? See Table 4.13. To analyze the data, the Minitab path is Stat/Basic Statistics/2 Proportions as shown in Fig. 4.36. (This is also available in the Two Proportions tab in the data file.) Then, step-by-step instructions are provided for this example.

Table 4.13   KIND Karz test of two proportions data

Category	Vendor 1	Vendor 2
Airbags with issues	200	180
Total airbags used	50,000	40,000

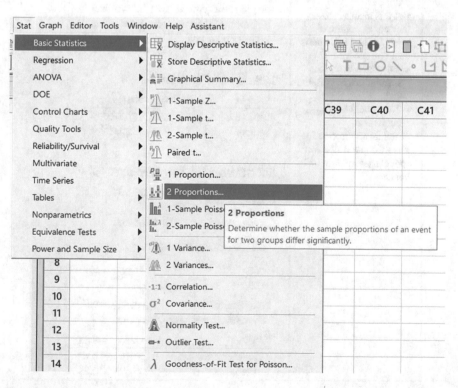

Fig. 4.36   Minitab path, test of two proportions

Step	Instructions	Illustration
1	Performing the analysis • Minitab path: Stat/basic statistics/2 proportions. • In the drop down at the top, select "summarized data". • Sample 1 will be for vendor 1, and sample 2 will be for vendor 2. • The "number of events" will be the numerator of the proportion, 200 and 180 for this example. • The "number of trials" will be the denominator of the proportion, 50,000 and 40,000 for this example. • Select "ok".	**Two-Sample Proportion** ✕ Summarized data ▼ Sample 1 Sample 2 Number of events: 200 180 Number of trials: 50000 40000 Select Options... Help OK Cancel
2	Interpreting the analysis • P-value is >0.05. • Therefore, you cannot conclude they are different. • But it is incorrect to say they equal. Failure to reject the null hypothesis could be due to: ◦ Too much variation. ◦ Limitations in sample size. ◦ They could be equivalent. • Practical application for this example: Look for root causes in both vendors.	**Test and CI for Two Proportions** **Method** p_1: proportion where Sample 1 = Event p_2: proportion where Sample 2 = Event Difference: $p_1 - p_2$ **Descriptive Statistics** Sample N Event Sample p Sample 1 50000 200 0.004000 Sample 2 40000 180 0.004500 **Estimation for Difference** 95% CI for Difference Difference -0.0005 (-0.001358, 0.000358) *CI based on normal approximation* **Test** Null hypothesis H_0: $p_1 - p_2 = 0$ Alternative hypothesis H_1: $p_1 - p_2 \neq 0$ Method Z-Value P-Value Normal approximation -1.14 0.253 Fisher's exact 0.255

Step	Instructions	Illustration
3	Test of two proportion watchouts • Low cell counts can be problematic—Use only Fisher's exact P-values if the number of events (not trials) is less than 5. • Small sample sizes can lead to high P-values. Try increasing your sample size (if possible) when the answer doesn't make sense.	**Test** Null hypothesis $\quad$ H_0: p_1 - p_2 = 0 Alternative hypothesis $\quad$ H_1: p_1 - p_2 ≠ 0 Method $\qquad$ Z-Value $\quad$ P-Value Normal approximation $\quad$ -1.14 $\quad$ 0.253 Fisher's exact $\qquad\qquad\qquad$ 0.255
4	*The project storyboard*: The following is an example how this information might appear in a project storyboard:	

Two Proportions: Conclusions $\quad$ Test and CI for Two Proportions

Method

• P value is > 0.05

• Therefore, can not conclude they are different

• But it is incorrect to say they are not different. Failure to reject the Null Hypothesis could be due to:
  • Too much variation
  • Limitations in sample size, or
  • They could be equivalent

• Practical application: Look for root causes in both vendors.

Measure

© Timothy Dean Blackburn 2020

4.6.3 Chi-Square Test of Multiple Proportions in Minitab

KIND Karz has been experiencing issues with brakes and is seeing an increase in warrantee work. Is one or more vehicle type worse than another? Can you tell by looking? (see Table 4.14).

Table 4.14 KIND Karz example for Chi-Sq test of multiple proportions

	Vehicle 1	Vehicle 2	Vehicle 3	Vehicle 4	Vehicle 5
Brake warrantee issues	200	20	50	50	0
No brake warrantee issues	749,800	649,980	249,950	49,950	450,000
Total vehicles produced	750,000	650,000	250,000	50,000	450,000
Percent defective	0.027%	0.003%	0.020%	0.100%	0.000%

In the prior example, we compared two proportions. But often there are more than two proportions as shown in the KIND Karz example. The Chi-Square test of multiple proportions can be used when this occurs. The Chi-Square value is calculated as shown in the following equation.

$$\chi^2 = \Sigma \frac{(\text{Observed} - \text{Expected})^2}{\text{Expected}}$$

This is quite easy to calculate in a spreadsheet or manually. The prior table can be expanded as shown in Table 4.15. Using the above equation, the Chi-Square value is 445.6. If the value exceeds the critical Chi-Square statistic at a significance of alpha = 0.05, the null hypothesis is rejected, and it is concluded at least one proportion is different than another.

The critical Chi-Square statistic is 9.49 for this example, which is less than the calculated Chi-Square statistic, so we can reject the null hypothesis. (This can be determined by using a standard Chi-Square table, or the Excel function = CHISQ. INV.RT(0.05, DF) with DF = number of proportions (n) −1 = 4 for this example.) It can also be calculated in Minitab using the Graph/Probability plot feature, with an output illustrated in Fig. 4.37.

By looking at the contribution to Chi-Square (243.4), it is clear Vehicle 4 is the most different and has the highest proportion of failures.

Also, this can be easily and quickly calculated in Minitab following the path Stat/Tables/Chi-Square Tabulation and Chi-Square and as shown in Fig. 4.38. Also, Minitab will provide additional helpful information. Below are step-by-step instructions.

Table 4.15 KIND Karz – Chi-Square calculations

Category	Vehicle 1	Vehicle 2	Vehicle 3	Vehicle 4	Vehicle 5	Total	Expected proportion
Brake warrantee issues	200	20	50	50	0	320	0.015%
No brake warrantee issues	749,800	649,980	249,950	49,950	450,000	2,149,680	99.985%
Expected brake issues	111.6	96.7	37.2	7.4	67.0	Total chi-Sq	Critical chi-Sq statistic
Chi-Sq value, brake issues	70.0	60.9	4.4	**243.4**	67.0	**445.6**	**9.49**
Expected no brake issues	749,888.4	649,903.3	249,962.8	49,992.6	449,933.0		
Total vehicles produced	750,000	650,000	250,000	50,000	450,000	2,150,000	
Percent defective	0.027%	0.003%	0.020%	**0.100%**	0.000%		

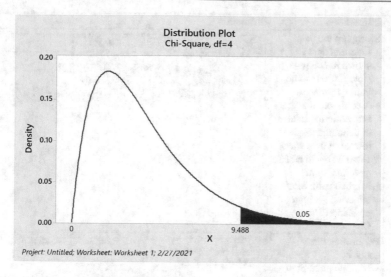

Fig. 4.37 Chi-Square critical statistic at DF = 4

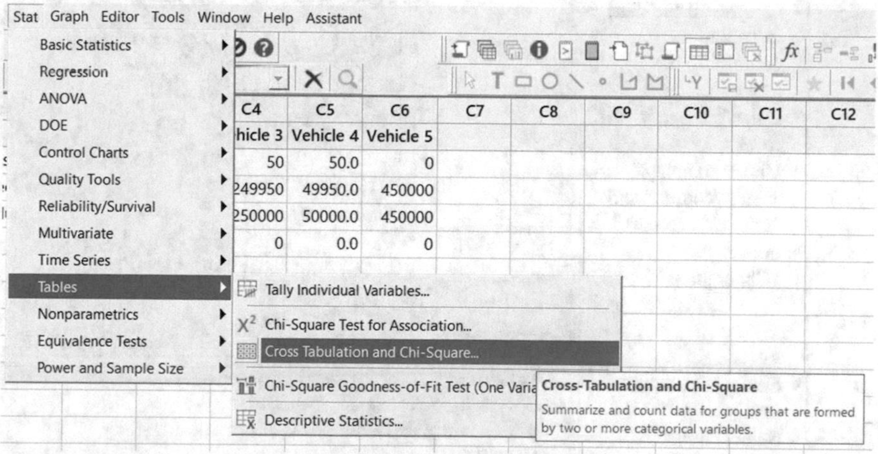

Fig. 4.38 Minitab path for Chi-Square test of multiple proportions

Step	Instructions	Illustration
1	Setting up the data • Enter the data as shown in Minitab. Note it needs to be in this format. • A common mistake is to enter the total or denominator in row 2 but note rows 1 and 2 should add up to the total. • Minitab path: Stat/tables/Chi-Square tabulation and Chi-Square.	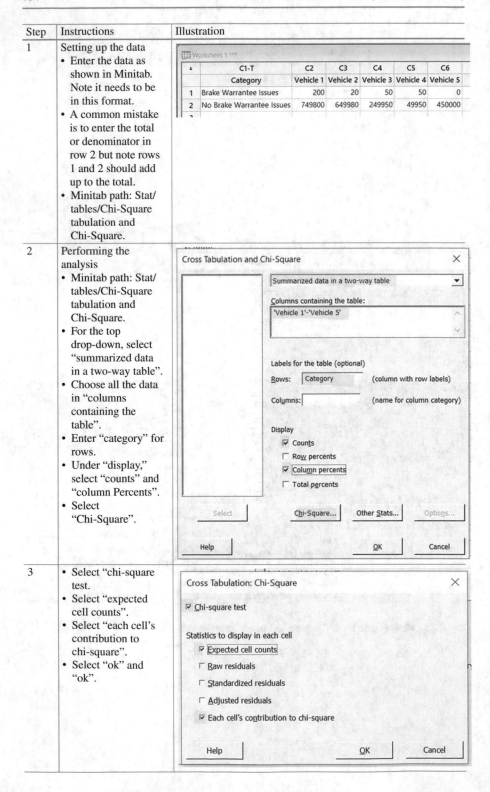
2	Performing the analysis • Minitab path: Stat/tables/Chi-Square tabulation and Chi-Square. • For the top drop-down, select "summarized data in a two-way table". • Choose all the data in "columns containing the table". • Enter "category" for rows. • Under "display," select "counts" and "column Percents". • Select "Chi-Square".	
3	• Select "chi-square test. • Select "expected cell counts". • Select "each cell's contribution to chi-square". • Select "ok" and "ok".	

Step	Instructions	Illustration
4	Interpreting the results • Note that the greatest contribution to chi-Sq is vehicle 4. • Note that P-value is <0.05. • Vehicle 4 is the most different (and highest) percent defective. • Practical application: Especially focus on vehicle 4 root cause analysis and corrective action.	
5	Watchouts • Significantly different counts of total (defective plus non-defective) can skew the results. • Low expected counts (<5) can yield an alert message. If so, look to increase sample size, combine categories to avoid low expected counts, or do Fisher's exact test (beyond the scope of this book). • Remember what contribution to chi-Sq is measuring: It is measuring the greatest contribution to chi-Sq. so sometimes the highest contribution to chi-Sq can be shared with least number of defects and greatest amount of defects.	
6	*The project storyboard*: The following is an example how this information might appear in a project storyboard: 	

References

Bothe, D. R. (1997). *Measuring process capability*. McGraw Hill.
Cintas, P., Almagro, L., & Labres, X.-M. (2012). Industrial statistics with minitab.
George, M. L., Rowlands, D., Price, M., & Maxey, J. (2005). *Lean six sigma pocket toolbook*. McGraw-Hill.
Gupta, H. C., Guttman, I., & Jayalath, K. P. (2020). *Statistics and probability with applications for engineers and scientists using MINITAB, R and JMP* (2nd ed.). Wiley.

Minitab Online Help. (n.d.). https://www.minitab.com/en-us/support/.
Ryan, B., Joiner, B. L., & Cryer, J. (2012). *Minitab handbook*. Cengage.
Sleeper, A. (2012). *Minitab DeMystified*. McGraw Hill.
Stagliano, A. A. (2004). *Six sigma advanced tools pocket guide*. McGraw Hill.

The Analyze Phase with Minitab Tools

5

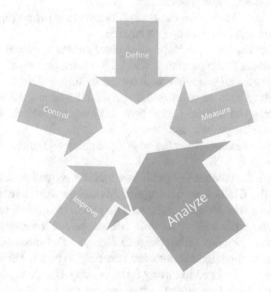

5.1   The Analyze Phase: An Overview

5.1.1   Introduction

Remember in the Analyze phase we will focus on identifying root causes (and verify the root causes with evidence). In this phase, we will work to achieve confidence that we have identified root causes with evidence that, when carried forward to Improve and resolved, will meet our project objectives. We still need to resist the temptation to focus on solutions, which should be done later in the Improve phase (except for obvious improvements needed). As with the prior phases, ensure the outcomes of the Analyze phase are included in a well-structured and easy to follow storyboard.

© The Author(s), under exclusive license to Springer Nature Switzerland AG 2022   107
T. D. Blackburn, *Six Sigma*, https://doi.org/10.1007/978-3-030-96213-5_5

Typical Analyze storyboard contents are as follows, which we will explore in more detail throughout this section:
- Questions and scope to carry forward from the Measure Phase.
- Identification of potential root causes.
- Root Cause Analysis visuals (fishbone or Causal Tree).
- Accept or reject based on evidence.
  - Process analysis.
  - Data analysis.
  - In-scope analysis.
- Possible: Outcomes of pilots or DOEs.
- Summary of the Analyze phase (verified and in-scope root causes).

As with each phase, there should be a clear transition between phases, avoiding ambiguity for the reader. For the KIND Karz case study, let's review the following conclusions from the Measure phase:
- Processes are not capable for the overall defect rate for airbags, airbag fabric tensile strength, and brake torsion resistance.
- Airbags and brakes have the highest incident counts for recalls and warrantee claims in the first 12 months after a new KIND Karz purchase.
- No difference between vendors was detected for airbag defects.
- Vehicle 4 tends to have more brake defects than the other vehicles.
- While the airbag test method was shown to be adequate as designed, it yielded low accuracy for the pinhole test.
- Airbag defects worsened after the new assembler equipment was installed at the vendor.

These provide a foundation to establish a focused scope for the Analyze phase. Continuing with the KIND Karz case study, this can be written in narrative form as follows: *The focus of the Analyze phase will be related to identifying root causes of airbag failures and brake caliper torsion failures, which account for 79.1% of all recalls and warrantee claims within the first 12 months of automobile purchase.*

This also leads to fundamental questions needing to be answered in the Analyze phase. For example, as related to airbag failures, why are we experiencing pinhole test failures? And why did the rate of airbag defects worsen after the new assembler was installed at the vendor? Also, we need to answer questions for the brake calipers as well, such as why are they failing? And why does vehicle 4 tend to have more brake defects than other vehicles?

To address these questions and eventually determine root causes, begin with a cause and effect analysis to identify potential root causes, leveraging clues gathered from the Measure phase.

---

[1] Note: If the P-value doesn't show in Normal Probability plot, go to Tools/Options/Linear Models/ Residual Plots (V18) or Files/Options/Linear Models/Residual Plots (V19–21) and check the box beside "Include Anderson-Darling test with normal plot" and rerun the analysis.

5.1.2 Cause and Effect Analysis

Here the question of "What are potential root causes?" is addressed. A common tool used for this is the fishbone diagram, which is also referred to as an Ishikawa diagram, or cause and effect (C&E) diagram (see Fig. 5.1). Notice that it looks like the bones of a fish, hence its name. (Although the fishbone diagrams herein were created in Microsoft © Visio, they can also be created in Minitab, Stat/Quality Tools/ Cause-and-Effect.)

A brief description of the problem is entered in the head of the fish. Then the large bones represent categories of potential root causes (not root causes themselves). These can be generically labeled from the six Ms., or Methods, Machines, Manpower (or People), Materials, Measurement, and Mother Nature (or Environment). The little bones represent potential root causes and use the 5 Whys method which will be described further. This approach is useful for brainstorming potential root causes with a team. Also, there could be multiple fishbone diagrams depending on the nature of the problems.

Start identifying the major categories (big bones) as described earlier. The standard categories are defined as follows:

- Machines: Tools, equipment, machinery, and facility/utility systems.
- Methods: Support processes, testing methods, procedures/SOPs.
- Materials: Raw materials, components, supplies.
- Mother Nature: Weather and environmental conditions and supporting systems (such as HVAC).
- Manpower: People issues, human error.
- Measurements: Measurements, manual and automated (could overlap with "Methods").

Then move to the smaller bones, brainstorming potential root causes. Keep drilling down and branching until the terminal root cause is identified and confirmed (or ruled out as a root cause). Often, five or less iterations are required to arrive at the

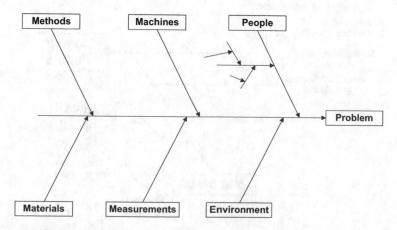

Fig. 5.1 Fishbone diagram

terminal root cause and is called 5 Why analysis. Knowing where to stop (and how to write the potential root causes) takes practice.

For the KIND Karz example, Albert started with a fishbone diagram as shown in Fig. 5.2 but quickly realized at least two fishbone diagrams would be needed (one for airbags and the other for the brake calipers). For the combined fishbone, notice a succinct and clear problem is written in the head of the fish, "Recalls and Warrantee Claims."

In actual practice, drawing a fishbone will be (and should be) somewhat busy and messy and would be difficult to read if shown in its completed state here. Therefore, it is simplified for illustration purposes.

The Materials large bone example is shown in Fig. 5.3. Note for this example the team identified 4) incorrect material thickness and 5) bad glue as potential root

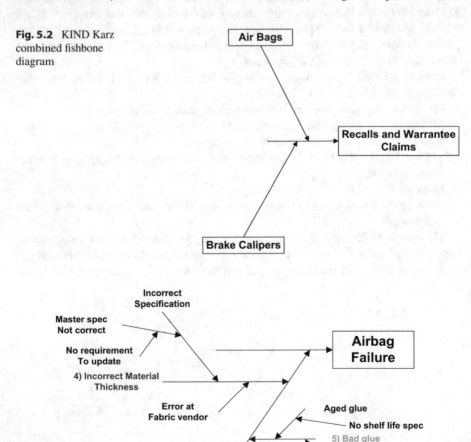

Fig. 5.2 KIND Karz combined fishbone diagram

Fig. 5.3 KIND Karz fishbone, detailed example

causes in the Materials category. The team kept drilling down until the potential root causes were identified. For example, for the incorrect material thickness, one terminal root cause was error at the fabric vendor. Another was no requirement to update specifications, after asking multiple whys: "Why is the material thickness incorrect?" Then, "why the incorrect specification?" Then, "Why is the master spec incorrect?" And then note the terminal potential root cause, "No requirement to update (the master spec)." See something similar for the bad glue potential root cause.

The team should retain the paper fishbone and notes, but for purposes of the storyboard, it would be unreadable. Typically, the fishbone is summarized as shown in Figs. 5.4 and 5.5. As the Analyze phase progresses, the potential terminal root

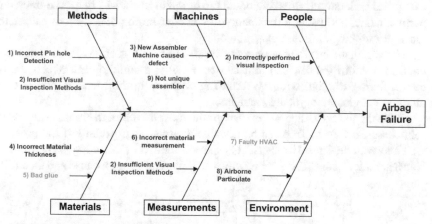

Legend: Red: Confirmed Root Cause; Blue: Out of scope; Green: Conformed not a root cause

Fig. 5.4   KIND Karz summary fishbone diagram, airbag failure

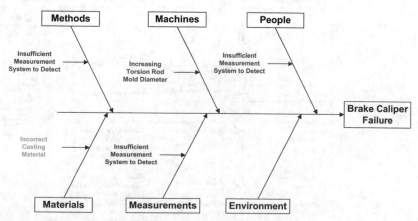

Legend: Red: Confirmed Root Cause; Blue: Out of scope; Green: Conformed not a root cause

Fig. 5.5   KIND Karz summary fishbone diagram, brake caliper failure

causes will need to be confirmed or discarded with evidence. It is helpful to color code which were confirmed, discarded, or deemed out of scope for the project. Make sure the items on the chart are readily and commonly understood. Also ensure the potential root causes can be identifiable later (e.g., use numbers or letters for easy cross reference).

There are common mistakes made when using a fishbone diagram. For example, the team might stop at one why, failing to drill down as far as practicable to a terminal root cause. Another is showing categories on little bones versus the root cause (e.g., "Glue" versus "Bad Glue"). Also, sometimes teams write a lack of a solution versus the root cause (e.g., "Need a new test method" versus "Faulty test method").

Sometimes fishbone diagrams are incomplete. While it is permissible to use other categories than the six Ms., only do so when the process is well known and categories understood. Breaking down the problem in six Ms. helps the brain compartmentalize, and usually the team will consider more potential root causes than when just looking at the problem singularly.

Another mistake is to vote on root causes—as noted prior, evidence is needed to confirm rule out (or discard) root causes. Finally, a common mistake is the root causes failing to align with the funneling exercise in the Measure phase, or not aligning with later storyboard elements.

While the fishbone diagram is perhaps the most used tool to brainstorm potential root causes, a Causal Tree can be a viable alternative, especially if the process is well known. The Causal Tree can also be used for 5 Whys. See Fig. 5.6 for a KIND Karz example. Causal Trees begin with a statement of the root cause and then follow

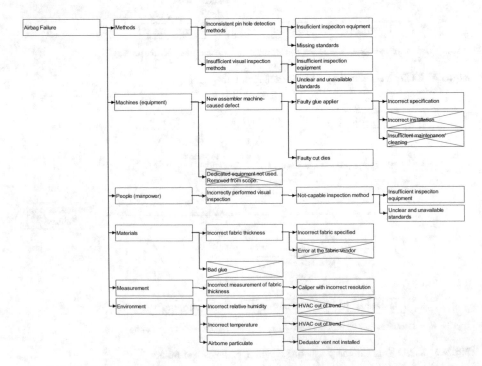

Fig. 5.6 KIND Karz Causal Tree example

a decomposition horizontally until the terminal root cause is determined. Note in this example, potential root causes that were determined not to be terminal root causes were crossed out.

5.1.3   Verifying or Discarding Root Causes: Process Analysis

As noted prior, the team should not vote on which root causes to carry forward. Instead, continue to follow a data-driven approach and verify or discard root causes only with evidence. The two methods for evidence include process analysis and data analysis. A combination of both might be required depending on the nature of the root cause. This data-driven rigor is a distinguishing factor for the Six Sigma approach. Practically, if the root causes are not confirmed, the problem might not be resolved.

Process analysis includes understanding the process, and mapping it as it is, versus how it should be or perceived to be. It requires walking the process with individuals familiar with the actual operation. It includes asking questions to get clues, such as, "When did it work properly, and then not? What changed? Is there another process or equipment that is working properly, and what is different?"

Process analysis also includes understanding first scientific principles (or what is physically possible). And it can take on a data review approach when statistical tools are not practical, such as reviewing records (alarms or excursions). It can also include analytical testing or deeper engineering domain analysis (e.g., advanced techniques for a specific discipline such as finite element analysis, etc.).

A key component of process analysis includes interviewing operators and subject matter experts (SMEs). Here are some practical tips for effectively interviewing. First receive permission (from their manager) and ask their manager to make them aware you will be interviewing them (to avoid a surprise). When interviewing, first reassure them it isn't a faultfinding exercise but a desire to solve the problem. However, avoid giving false assurances or promises of not impact to the individual, as you likely don't have the authority to make a discipline decision.

After you have introduced yourself, and hopefully put them at ease, ask for their help. Most people want to help (within reason). Then, ask open-ended questions. This can include,

- Tell me what happened?
- Was anything different or unusual at the time?
- Has this happened before?
- What can you tell me about the issue?
- What do you think the causes are?

Asking open-ended questions can usually lead to better and more information than close-ended questions. Remember to ask them to show you versus explain. Look for clue words where root causes might be (such as tricky, sometimes, adjust, experience, overtime, tweak). For example, an operator might say, "This adjustment is tricky and requires experience." That would be a situation where you should continue to ask questions, as it might represent variation in the process that led to the event.

As an example, for KIND Karz, the team brainstormed earlier that a potential root cause for the airbag failure could be particulate in the seal zone, where the fabric adheres to the airbag housing. In the past, this has been known to lead to failure. The team discussed potential sources of particulate, such as failure of the differential pressure controls in the manufacturing room that could have allowed particulate from an adjacent dusty operation to enter the room. Also, there could be particulate generating materials inside the manufacturing room.

As part of process analysis, Albert walked the process to understand the steps in sealing. This included interviews with personnel, with a variety of open-ended questions, along with a direct question as to whether particulate generating materials were used in the room. Albert also reviewed alarm and differential pressure records from the building management system and saw no deviations during the time of the faulty airbag manufacturing. He was unable to identify any concerns during the time of manufacture that would have generated excessive particulate. But to be certain, he asked the laboratory to compare the seal zone of the failed airbags to examples that had not failed. The results came back negative, as there was not a demonstrable difference between good and failed airbags (see Fig. 5.7).

So, Albert discarded the potential root cause of particulate. He summarized this for the team and in the storyboard as follows:

- The manufacturing suite is positively pressurized for adjacent areas to avoid the concerns of airborne particulate affecting the airbag seal.

Fig. 5.7 Airbag seal images

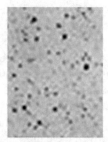

Good Airbag Seal Image

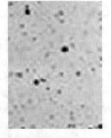

Failed Airbag Seal Image

- This is continuously monitored by the Building Management System (BMS). All records were reviewed during the fabrication period of the failed airbags, but there were no alarms indicated.
- Preventive maintenance records were also reviewed, and there were no concerns noted during the period.
- Interviewed personnel—no presence of particulate generating materials are permitted in the space (and no exceptions were noted during the affected period).
- Finally, bag seal zones with failures were compared under a microscope to those that did not fail, and there was no visibly appreciable difference.
- Therefore, this root cause is ruled out.

Another example of process analysis for the KIND Karz case study is related to why vehicle 4 has more caliper failures than the other vehicles. (Also, it would be interesting to see why vehicle 5 had none.) After reviewing the vehicle types, Albert discovered vehicle 4 was a pickup truck and had used the same calipers as small sedans. After further review and discussions, he determined this was due to a design error. The incorrect design was selected from the truck from a set of common designs for cars. However, to confirm the incorrect caliper was used, the structural engineering team performed finite element analysis which indicated the stresses from the heaver vehicle exceeded the capacity of the torsion rod (even at a full diameter), leading to greater failures. Here in the Analyze phase (versus Improve), the solution was already clear—use the correct design going forward. In this case, we don't have to wait to the later phase, especially when critical or life threating. KIND Karz immediately initiated a recall for the calipers to be replaced.

The previous first example was when a root cause was discarded or rejected and the second when it was accepted using process analysis. In some cases, a third option can be considered—descoping the potential root cause. This is sometimes necessary when available time or budget do not permit resolving the root cause. For example, funding to resolve the root cause might need to be included in another budget cycle. Or, the scope could be too extensive, requiring another project. However, when descoping a potential root cause, make sure it isn't for convenience (or because the evidence is difficult to determine). Also make sure it will not result in an environmental, health, or safety risk. Finally, do a pro forma analysis to estimate the improvement and whether the project goal will still be achieved if the potential root cause is descoped.

While in these examples process analysis was the predominate methods used to accept or discard potential root causes, data analysis should be used whenever possible and often is used in combination with process analysis.

5.1.4   Verifying or Discarding Root Causes: Data Analysis

Data analysis includes visualization and various statistical techniques. While there is a myriad of statistical methods used, some of the more common ones will be illustrated in this section and in companion detailed sections. These include but are not necessarily limited to the following:

- Regression: $Y = f(X)$.
- Two Sample T Tests, ANOVA: Comparing means of different groups.
- Paired T Tests: Comparing paired data.
- Test of proportions—are proportions different? Detailed instructions on how to perform test of proportions is included in Sect. 4.6 and not repeated in this section.
- DOE (Design of Experiment): Which factors are significant? How to achieve the goal for your Y?

The following will include examples of how the above statistical methods might be applied in a situation like KIND Karz project, to confirm or discard root causes. It will be presented from a team member or stakeholder point of view, and details on how to run the analysis for a Six Sigma practitioner will follow in companion sections. The examples will include and conclude remaining potential root causes.

But first, let's review the importance of and practical applications of piloting. Then regression will be reviewed, followed by the remaining statistical methods.

5.1.5 Piloting

First, remember from the Measure phase the brake caliper torsion test device had distinct categories less than 5, which led to potential root cause 1b. Remember Albert performed a Gage R&R in the Measure phase which led to this conclusion. This is an example of a pilot, which can be helpful in the Analyze phase to confirm assumptions of root causes.

After reviewing the test device with the test equipment vendor, Albert discovered it was designed to enable that level of resolution. However, the test equipment vendor did offer a different test device model that was designed to a higher resolution, which KIND Karz piloted. Such pilots can be helpful in the Analyze phase to confirm assumptions of root causes. A Gage R&R was repeated after the pilot, and the results are shown in Fig. 5.8. Note there are now seven distinct categories, and the % study variation is less than the maximum of 30% recommended. This root cause is therefore confirmed, and the solution will be carried forward to the Improve phase for further evaluation.

5.1.6 Data Analysis Example: Regression

A common statistical method used in the Analyze phase is regression analysis (although it is not limited just to the Analyze phase). To illustrate this method, consider potential root cause 2b, which was related to the brake caliper torsion rod issue, and specifically a concern with the mold size increasing over time. Recall from the control chart lecture, the defects started to occur after the mold diameter increased to ¼" diameter. The root cause was expected to be related to failure of the deslagger device to properly remove slag from the reusable mold during caliper casting. The mold diameter had buildup over time that led to a smaller diameter torsion rod in the caliper. This trend was confirmed using the run chart (in a prior section). Therefore, the reduction in the patented caliper torsion rod component would clearly reduce torsion resistance.

Gage R&R

Variance Components

Source	VarComp	%Contribution (of VarComp)
Total Gage R&R	0.09143	7.76
Repeatability	0.03997	3.39
Reproducibility	0.05146	4.37
Assessor	0.05146	4.37
Part-To-Part	1.08645	92.24
Total Variation	1.17788	100.00

Gage Evaluation

Source	StdDev (SD)	Study Var (6 × SD)	%Study Var (%SV)
Total Gage R&R	0.30237	1.81423	18.42
Repeatability	0.19993	1.19960	4.37
Reproducibility	0.22684	1.36103	20.90
Assessor	0.22684	1.36103	20.90
Part-To-Part	1.04233	6.25396	96.04
Total Variation	1.08530	6.51180	100.00

Number of Distinct Categories = 7

Fig. 5.8 KIND Karz: piloted brake caliper test device, Gage R&R output

After process analysis (including reviewing the device and discussing with the operators), Albert identified potential factors that could be leading to a drift over time. These factors included abrasion settings, lubricant pH, ambient temperature, and relative humidity. Albert first used a data visualization technique in Minitab, called a "Matrix Plot" as shown in Fig. 5.9. Details on how to generate this plot are included in Sect. 5.2: "Regression Analysis – Minitab Methods and Analysis Detail."

Note the response variable (Caliper Mold Diameter) is on the bottom row of the chart. The columns are the four factors identified. Where each intercepts the last row, we should review to determine if there is an apparent relationship with mold diameter, indicated by a sloping line and points massing around the line. In this case, it appears there is a relationship between the deslagger settings and lubricant pH. However, later evaluations determined that pH is not a directly contributing factor, changed from the abrasion heat (as affected by the abrasion settings, evidenced in the panel and the sloping line where Lub pH and Abrasion Setting intersect).

To confirm the apparent relationship between abrasion settings and mold diameter, Albert performed regression analysis using the Fitted Line Plot tool in Minitab (see Fig. 5.10). He reviewed historical data (same as with the Matrix Plot) and plotted the relationship between the abrasion settings (X axis) and the mold diameter (Y axis). Then using Minitab, he fitted a line that best represented the plotted data. Note the R-Sq value (which is the correlation coefficient squared) indicates the percent of

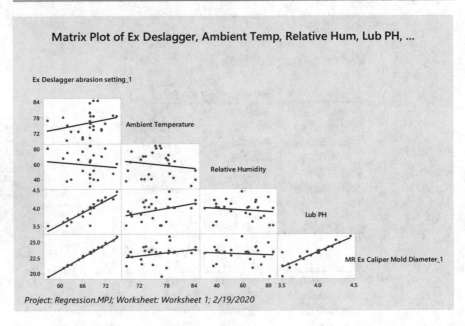

Fig. 5.9   KIND Karz Caliper Mold Diameter Matrix Plot

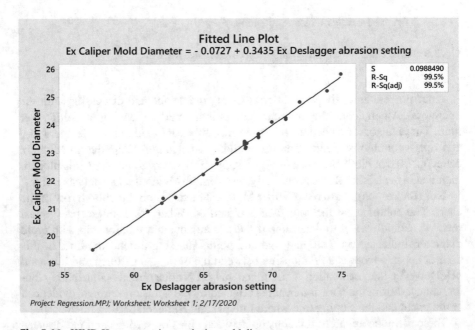

Fig. 5.10   KIND Karz regression analysis, mold diameter

variation in the Y (Mold Diameter) as predicted by the abrasion setting to be 99.5%. This is very high and suggests a relationship between the two factors. Details on how to perform the regression analysis is included in Sect. 5.2.

5.1.7   Data Analysis Example: Two Sample T Test

Another common statistical method used in the Analyze phase (but not necessarily limited to this phase) is the Two Sample T Test. The level of detail provided will be typical for a team member or stakeholder. For details on how to perform a Two Sample T Test, see Sect. 5.3.

To illustrate how this statistical technique might be used, remember potential root cause 3a, which is related to the airbag defects worsening after the new assembler was installed at the vendor's manufacturing plant. To confirm whether the airbag defect rates worsened after the new equipment was installed, Albert gathered data before and after the time of the installation and then performed a Two Sample T Test which he shared with the team. The Minitab output data is shown in Fig. 5.11.

Two-Sample T-Test and CI: Defects per Airbag_1, Shift_1

Method

μ_1: mean of Defects per Airbag_1 when Shift_1 = After
μ_2: mean of Defects per Airbag_1 when Shift_1 = Before
Difference: $\mu_1 - \mu_2$

Equal variances are not assumed for this analysis.

Descriptive Statistics: Defects per Airbag_1

Shift_1	N	Mean	StDev	SE Mean
After	16	0.3702	0.0299	0.0075
Before	28	0.20152	0.00943	0.0018

Estimation for Difference

Difference	95% CI for Difference
0.16865	(0.15235, 0.18494)

Test

Null hypothesis	H_0: $\mu_1 - \mu_2 = 0$
Alternative hypothesis	H_1: $\mu_1 - \mu_2 \neq 0$

T-Value	DF	P-Value
21.94	16	0.000

Fig. 5.11   KIND Karz, airbag assembler, Two Sample T Test

The P-value is less than alpha 0.05, so the null hypothesis that the average defect rates are the same before and after the equipment was installed is rejected, and the alternative hypothesis that there is a difference is accepted. Specifically, the average defect rate with the old equipment was 20.1%, and the defect rate with the new equipment was 37%. Clearly, something changed that needs to be evaluated further. This is an example where a root cause is partially confirmed, but further evaluation is needed which will be addressed later. But first, let's look at example of a similar analysis method but for paired data, the paired T Test.

5.1.8   Data Analysis Example: Paired T Test

Another common statistical method used in the Analyze phase (also not limited to this phase however) is the paired T Test. The level of detail provided will be typical for a team member or stakeholder. For details on how to perform a paired T Test, see Sect. 5.4. The difference between a paired T Test and a Two Sample T Test is the data is paired, as will be illustrated in the following KIND Karz example.

Remember from before potential root cause 61, which was a concern that the incorrect material measurement might have been taken at the fabric vendor, allowing a different thickness to have been shipped. Albert visited the fabric vendor's factory (which took some negotiation as it was a vendor to a vendor). Remember KIND Karz sourced the airbags to a third party.

After visiting the factory, and performing a process analysis, Albert discovered that two different tools were used to calibrate fabric thickness settings on the machine that produced the fabric (referred to as Device A and Device B). These were measurement tools that measured the thickness and based on the thickness automatically adjusted the machine. After further evaluation, Albert noticed that Device A was used on fabric that had failures and Device B was used on fabric that had no failures. Could it be something was different with the calibration tools that led to failures? So, Albert decided to do a statistical analysis—he collected several swatches of fabric and measured thickness using both calibration tools. Then, he performed a paired T Test to see if there was a statistically significant difference between Device A and B.

The Minitab output and results are shown in Fig. 5.12. For paired T Test, the difference in readings for each fabric swath are calculated. The null hypothesis is the differences are zero, or there is no difference between Device A and B. However, the P-value is less than alpha 0.05, so the null hypothesis is rejected, and it is assumed the difference is not zero. That is, there is a statistically significant difference in the fabric thickness measurements between Device A and B.

This led to further evaluation of the devices, including comparing to a standard. It was concluded that Device B gives a false high reading, which led to an over-calibration of the fabric machine, resulting in thinner fabric. Evidence is then provided to support and confirm the root cause of thinner fabric.

Paired T-Test and CI: Device A, Device B

Descriptive Statistics

Sample	N	Mean	StDev	SE Mean
Device A	10	0.030246	0.000929	0.000294
Device B	10	0.033271	0.001022	0.000323

Estimation for Paired Difference

Mean	StDev	SE Mean	95% CI for μ_difference
-0.003025	0.000093	0.000029	(-0.003091, -0.002958)

μ_difference: mean of (Device A - Device B)

Test

Null hypothesis	H_0: μ_difference $= 0$
Alternative hypothesis	H_1: μ_difference $\neq 0$

T-Value	P-Value
-102.96	0.000

Boxplot of Differences

Fig. 5.12 KIND Karz: Device A and B airbag fabric calibration tools

So far, we have reviewed and contrasted applications when we want to evaluate whether two samples have different means. What if we have more than two samples? The next section will illustrate an example for KIND Karz using ANOVA.

5.1.9 Data Analysis Example: ANOVA, ANOM

Another common statistical method used in the Analyze phase (also not limited to this phase however) is ANOVA. This is needed when we want to determine whether there is a difference between means of more than two samples. The level of detail provided in the following will be typical for a team member or stakeholder. For details on how to perform ANOVA, see Sect. 5.5.

Remember from before potential root cause 3b, which was a concern that the cast iron composite material used for the calipers was an incorrect material, which might have led to a lower tensile strength. KIND Karz had developed its own special alloy of cast iron. There were four different lots of raw material used in the time period of the most failures. The primary measure of cast iron quality for this special alloy is tensile strength. An average tensile strength of 414 Mpa is considered acceptable.

Lot 1 failed calipers	Lot 2	Lot 3	Lot 4
409.594	412.598	413.595	416.909
406.018	424.459	417.886	404.425
439.926	419.445	423.270	418.407
391.406	398.386	420.456	395.683
403.661	408.199	409.561	419.951

Fig. 5.13   KIND Karz cast iron tensile strength

One-way ANOVA: Lot 1 failed calipers, Lot 2, Lot 3, Lot 4

Method

Null hypothesis	All means are equal
Alternative hypothesis	Not all means are equal
Significance level	$\alpha = 0.05$

Equal variances were assumed for the analysis.

Factor Information

Factor	Levels	Values
Factor	4	Lot 1 failed calipers, Lot 2, Lot 3, Lot 4

Analysis of Variance

Source	DF	Adj SS	Adj MS	F-Value	P-Value
Factor	3	385.3	128.4	1.11	0.346
Error	116	13375.4	115.3		
Total	119	13760.7			

Fig. 5.14   KIND Karz cast iron tensile strength ANOVA

Interestingly, Albert discovered that all the failed torsion rods were from Lot 1. The team became excited, and someone said, "That's it! It must be something to do with that lot." But Albert smiled, "Slow down – let's see what the data tells us."

He had test coupons made from each of the lots, and each was tested for yield (maximum tensile strength). The results (partially) are shown in Fig. 5.13 to illustrate the data construct necessary to perform this analysis.

Then he ran analysis of variance (ANOVA), which indicated there was insufficient evidence to conclude one mean was different than another. Note the P-value is greater than alpha 0.05, which indicates there is insufficient evidence to reject the null hypothesis. Therefore, based on ANOVA, he could not conclude Lot 1 was different than the other lots (Fig. 5.14).

Similarly, he ran analysis of means (ANOM), which also indicated there was insufficient evidence to conclude one mean was different than the mean of the means (see Fig. 5.15). Note that there are no points outside the decision limits.

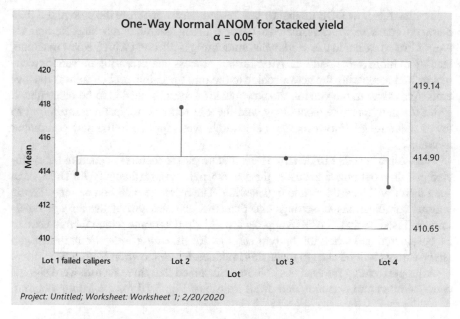

Fig. 5.15   KIND Karz cast iron tensile strength, ANOM

Based on ANOM and ANOVA results, the team could not conclude the batch of cast iron used for defective calipers had a different tensile strength from cast iron used for non-defective calipers. Therefore, this root cause was discarded. As a clarifying note, when we fail to reject the null hypothesis, we cannot say the means are the same, but rather we should say we cannot conclude they are different. This will be explained further Sect. 5.5.

The details on how to perform ANOVA (and a nonparametric alternative) and ANOM, as well as how to check for statistical analysis assumptions, are included in Sect. 5.5. But next let's look at another common analysis method used in the Analyze phase—Design of Experiment (DOE).

5.1.10 Data Analysis Example: Design of Experiment (DOE)

Another common statistical method used in the Analyze phase (also not limited to this phase however) is Design of Experiment (DOE). Often, as was shown in most of the prior examples, historical data is frequently available for analysis. However, available historical data might not have sufficient ranges or might be missing key potential predictors or causes. DOE will allow us, after further funneling and identifying likely predictors, to modify the process by setting different levels and generate data to better identify significant factors and more optimal settings. The level of detail provided in this section will be typical for a team member or stakeholder. For details on how to perform fundamental DOE, (step-by-step in Minitab for the examples shown in the following), see Sect. 5.6.

To illustrate how DOE might be used, remember from before the potential KIND Karz root cause 3a, which is related to the new airbag assembler machine. Remember we confirmed earlier in this section that something is different with the new machine, leading to more airbag failures. After further process analysis with the vendor technician, and comparing the new machine to the old, a possible root cause was discovered. According to the vendor, the new machine settings needed to be recalculated given the new machine technology, and the old settings might be contributing to lower seal strength. However, the old settings were applied to the new equipment as well.

Next Albert needed to determine what settings (or factors) needed to be evaluated and then optimized to ensure the seal strength was sufficient. With DOE, each such factor will have a low and high setting. The machine will then be run with the various combinations of settings and then the seal strength of the airbags determined. Based on that, the DOE approach will help determine which factors need to be considered and what are the best settings for maximum seal strength. Minitab analysis tools will be used to make this much easier as shown in Sect. 5.6.

After performing the analysis, Albert determined that three factors were significant—temperature, vacuum, and dwell time. See Fig. 5.16 for the Minitab output.

Analysis of Variance

Source	DF	Adj SS	Adj MS	F-Value	P-Value
Model	4	353816	88454	3719.30	0.000
Linear	3	353036	117679	4948.14	0.000
Temp	1	154846	154846	6510.95	0.000
Vacuum	1	136765	136765	5750.66	0.000
Time	1	61425	61425	2582.80	0.000
2-Way Interactions	1	780	780	32.80	0.000
Temp*Vacuum	1	780	780	32.80	0.000
Error	27	642	24		
Lack-of-Fit	11	2	0	0.00	1.000
Pure Error	16	640	40		
Total	31	354458			

Model Summary

S	R-sq	R-sq(adj)	R-sq(pred)
4.87672	99.82%	99.79%	99.75%

Coded Coefficients

Term	Effect	Coef	SE Coef	T-Value	P-Value	VIF
Constant		1432.25	0.86	1661.37	0.000	
Temp	139.125	69.563	0.862	80.69	0.000	1.00
Vacuum	130.750	65.375	0.862	75.83	0.000	1.00
Time	87.625	43.812	0.862	50.82	0.000	1.00
Temp*Vacuum	9.875	4.937	0.862	5.73	0.000	1.00

Regression Equation in Uncoded Units

Airbag Seal Strength with Inter = 92.9 + 4.981 Temp - 2.84 Vacuum + 0.8762 Time
                                  + 0.04937 Temp*Vacuum

Fig. 5.16 KIND Karz new assembler DOE Minitab output

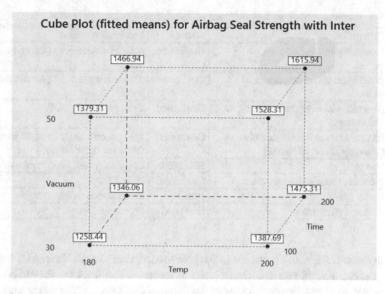

Fig. 5.17   KIND Karz new airbag assembler DOE cube plot

Note that the P-values for each of the factors are less than alpha 0.05, indicating the factors are significant. Also, he determined there was an interaction between temperature and vacuum, meaning the setting of one depended on the setting of another (see the two-way interaction low P-value in Fig. 5.16). Further note the Minitab analysis output provided a predictive model or equation and had a very good R-sq(adj) value (99.79%), indicating the model predicts well.

After identifying which factors were significant, next Albert determined the new settings to ensure a maximum seal strength. Given there are three factors significant, he was able to visualize this for the team using the cube plot as shown in Fig. 5.17. Note that the highest seal strength is 1645.94 when temperature is 200, vacuum is 50, and time is 200. These will be the new standard settings for the new assembler machine to maximize seal strength.

5.1.11 Summary Root Cause Tables

It is helpful to include in the storyboard a table of potential root causes identified in the initial fishbone exercise, along with their resolution. In these tables, show all the potential root causes and status of resolution (confirmed, discarded, or descoped). Also include the verification method. The root causes in the tables should clearly link back to the fishbone exercise, as well as process and data analysis that were used to confirm or rule out the root causes. See Table 5.1 as an example for the airbag root causes for KIND Karz. A similar table was provided for the brake calipers.

Root cause summary tables are essential as we proceed to the Improve phase, where identification of solutions and implementation will occur. Include these in

Table 5.1 Root cause analysis summary and verification method, airbag defects

ID	Description	Status	Verification method
1a	Incorrect pinhole detection	Confirmed	Attribute agreement (see measure phase)
2a	Insufficient visual inspection methods	Confirmed	Attribute agreement (see measure phase)
3a	New assembler machine defects; low seal strength	Confirmed	2 sample T test, DOE
4a	Incorrect fabric tensile strength (incorrect fabric thickness)	Confirmed	Ppk; process analysis (used incorrect fabric for these airbags)
5a	Bad glue	Excluded	Laboratory test; confirmed expiry
6a	Incorrect material measurement	Confirmed	Paired T test
7a	Faulty HVAC	Possible	Temperature trends
8a	Airborne particulate	Excluded	Process analysis

our storyboard. As with all phases in the storyboard, don't make the reader work at it to understand what you did. Don't write to impress. The findings should be comprehensible to a reasonably informed audience. Maintain a clear story and clear thread connecting each DMAIC phase throughout the storyboard. That is, learnings in the Measure phase should carry forward to Analyze and then into Improve.

For KIND Karz example, Albert has now addressed the other questions needing to be answered in the Analyze phase.

• What are my verified root causes?
• What evidence do I need to reject or accept the root causes?
• Do I have sufficient process analysis or data analysis (e.g., statistical significance) to proceed to the Improve phase?

But there is one more question to answer, which is, "What are my conclusions from the Analyze phase, and which root causes am I carrying forward to the Improve phase?" Albert summarized these in the storyboard as follows:

1. Airbag.
    a. Test method is insufficient to detect pinhole leaks consistently.
    b. Device B gives incorrect fabric thickness reading at the vendor.
    c. High settings on temperature, vacuum, and dwell time impact seal strength.
    d. Fabric was exposed to excessive temperature in the warehouse.
    e. Incorrect fabric was shipped.
2. Brake caliper.
    a. Caliper used to measure torsion rod diameter doesn't have the proper resolution.
    b. The deslagger cleaning device settings are causing the mold of the torsion rod to decrease in diameter over time.
    c. The incorrect caliper was installed on vehicle 4 (pickup trucks).

It is now finally time to move to the Improve phase. The KIND Karz team broke out into an applause. They were ready for the next phase and eager to proceed.

But first, you might want to read the following sections that provide step-by-step instructions on how to do the analysis we have covered so far in the Analyze phase.

5.2   Regression Analysis: Minitab Methods and Analysis Detail

5.2.1   Regression Overview

Regression analysis is one of the best actualizations of $Y = f(X)$, where Y is the response variable (also called a *dependent variable*) and the X's are the predictors (also called the *independent variables*). While regression is often used in the Analyze phase as illustrated here, it is not limited to this phase.

For a single X and Y, this can be easily illustrated as a graph with the vertical axis being the Y and horizontal axis being the X. Linear regression assumes that a straight line best describes the relationship between X and Y. For example, consider the following data in Table 5.2.

When plotted, see a line that best fits the variation in the actual points (see Fig. 5.18). Regression analysis uses a method called *sum of the least squares*. To understand this, first note the differences between the actual data points and the point on the line that would be predicted at a given value of X. This difference is called a *residual*. In regression, each of the residuals is squared and then added together. The line position that has the lowest sum of these squared residuals is plotted as shown. From this line, we can determine an equation that is $Y = f(X)$. In this example, the equation is as follows:

$$Y = f(X) \text{ takes the form of } Y = a + bX, \text{ and for this example}$$
$$Y = 0.6955 + 0.09701\,X$$

where
- Y is the response variable.
- a = 0.6955, the Y intercept, or where the line crosses the Y axis when X = 0,
- b = 0.09701, the slope term, or the coefficient that is multiplied by X in the equation.

  And the equation can be used to predict Y at a given X. For example, when X = 50:
- $Y = 0.6955 + 0.09701\,X = 0.6955 + 0.09701(50) = 5.5460.$

Table 5.2 Sample regression data

↲	C1	C2
	Y	X
1	1	9.1219
2	2	20.6789
3	3	21.0773
4	4	33.0761
5	5	62.5069
6	6	43.8716
7	7	47.2866
8	8	76.1133

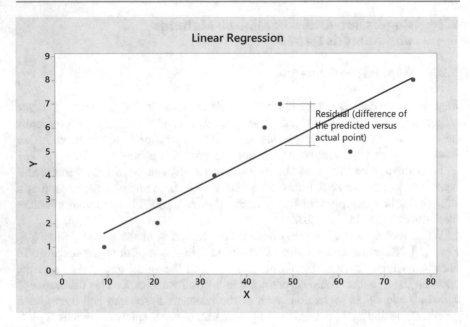

Fig. 5.18 Sample linear regression

5.2.2   Assumptions for Linear Regression and Key Data Interpretations

The following are assumptions and important key interpretations from regression analysis (which will be illustrated in the next example):

- The residuals must be normally distributed.
- There should be a random pattern of the residuals versus the predicted (or fitted) values. The residuals versus the order of observation should also be a random pattern.
- The b slope coefficient P-value should be less than 0.05 to be considered statistically significant (i.e., to be considered different than zero. A slope of zero would be no slope and a horizontal line). For single regression, this will be indicated as the overall *regression* P-value, but for multiple regression, the b_n P-values for each X_n will be provided.
- The R-Sq value will be between 0 and 1. Ideally, the R-Sq value should be closer to 1 to consider the factor(s) or X's in the model sufficient to predict Y. This measures the fractional proportion of variation in Y predicted by the variation in X. When there are multiple predictors (X's), use R-squared adjusted—this is to compensate for the effect of increasing R-Sq when X's are added. In practice, regression models might be considered useful with R-Sq values as low as 0.30 depending on the application but with 0.70 to 0.90 more desirable (especially when the equation will be used to predict Y versus just identifying significant X factors alone).
- While the X might predict Y, it cannot necessarily be assumed to be the cause. For example, people might tend to get sunburn more frequently when they eat ice

cream. But ice cream isn't the cause—people tend to eat more ice cream in the summer when they are outdoors.

- Only predict within the range of the data. Without knowing the wider range, it cannot be assumed the linear trend line continues. In the above example, X ranged from 9.1219 to 76.113, where the linear fit was apparent. But what if an X of 90 was used to predict Y? Would the equation still apply? Or would the line curve upward or downward beyond that point? In such a case, the regression equation generally should not be used. Instead, seek to gather a wider range of data that encompasses the full range of X's and reevaluate.

5.2.3 Single Linear Regression

To illustrate single linear regression, the following is an example from KIND Karz. The torsion rods were an added feature to brake calipers from KIND Karz innovation, which allowed for smoother braking. But some have failed and had the opposite effect, resulting in the recalls. Calipers (with their rods) are cast into reusable molds using a patented alloy, requiring the removal of slag periodically. Slag is a by-product of casting. A piece of equipment called a *Deslagger* is used to remove the slag, specially made for the new type of alloy. However, if too much material is removed, the mold will be damaged. If not enough is removed, the hole will become smaller in the mold, and the torsion rod will subsequently have a reduced diameter (which is what KIND Karz has observed). See Fig. 5.19 for a hand-drawn sketch Albert made in his notebook of the mold and the deslagging process.

The specification is to have at least 80 N-m torsion resistance in the torsion rods, which cannot be achieved with a small diameter. There is a concern the deslagger device abrasion force settings might not be correctly set. So, the project leader sought to determine if there is a relationship between the mold diameter and the abrasion force settings.

To follow along in the analysis, copy and paste the data from the "Regression" tab in the available Excel data spreadsheet into Minitab or recreate from the Appendix. The Minitab path to perform the regression is Stat/Regression/Fitted Line Plot. This is also shown in Fig. 5.20. Then detailed step-by-step instructions are provided along with key assumptions and considerations when performing single linear regression.

Fig. 5.19 Caliper mold deslagging process

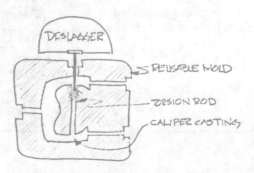

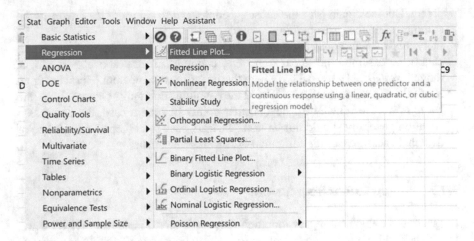

Fig. 5.20 Single linear regression Minitab path

Step	Instructions	Illustration
1	• From the data spreadsheet, go to the regression tab and copy and paste into Minitab first two columns of data. • Note the pairwise deslagger abrasion force settings. • Is there a relationship between the abrasion settings and mold diameter? • Is it possible to determine this by just observing the data? • Next step will be to do a regression analysis in Minitab.	<table><tr><th>B</th><th>C</th></tr><tr><td>Ex Caliper Mold Diameter</td><td>Ex Deslagger abrasion setting</td></tr><tr><td>19.50</td><td>57.00</td></tr><tr><td>20.90</td><td>61.00</td></tr><tr><td>21.20</td><td>62.00</td></tr><tr><td>21.36</td><td>62.10</td></tr><tr><td>22.22</td><td>65.00</td></tr></table> *Note: A portion of the data is shown only for illustration purposes. See the full dataset*
2	Minitab path for a fitted line plot: • Stat/regression/fitted line plot.	

Step	Instructions	Illustration
3	• For the response, select the "ex caliper Mold diameter". • For the predictor, select the "ex Deslagger abrasion setting". • Select "linear." note there are two other models available if a better fit is nonlinear. • Select "graphs".	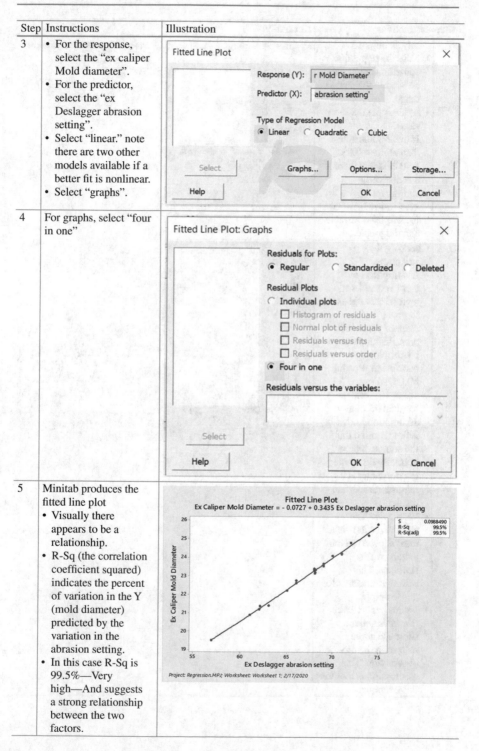
4	For graphs, select "four in one"	
5	Minitab produces the fitted line plot • Visually there appears to be a relationship. • R-Sq (the correlation coefficient squared) indicates the percent of variation in the Y (mold diameter) predicted by the variation in the abrasion setting. • In this case R-Sq is 99.5%—Very high—And suggests a strong relationship between the two factors.	

Step	Instructions	Illustration
6	Analysis output • Note the regression equation should only be used in the data range. • Again, note the R-sq value. • If the P-value is less than alpha = 0.05; conclude that there is a statistically significant association between the response variable and the predictor.	**Regression Analysis: Ex Caliper Mold Diameter versus Ex … ion setting** The regression equation is Ex Caliper Mold Diameter = - 0.0727 + 0.3435 Ex Deslagger abrasion setting **Model Summary** S　　　　R-sq　R-sq(adj) 0.0988490　99.51%　　99.49% **Analysis of Variance** Source　　　DF　SS　　　MS　　　F　　　P Regression　1　47.6382　47.6382　4875.41　0.000 Error　　　24　0.2345　0.0098 Total　　　25　47.8727
7	Checking the assumptions • Assumptions must be met to consider the regression valid as follows: • *Normal probability plot*: Residuals must be normally distributed. P-value >0.05, so this example meets the residual normality assumption.[1] note: It is not required that the source data be normally distributed, but normality is a requirement for the residuals. • Versus *fits*: The predicted (fits) values must show a random pattern versus residuals. This assumption is met for this example. • Versus *order*: The residuals should show a random pattern in the order of observation. This assumption is met for this example.	 Residual Plots for Ex Caliper Mold Diameter Project: Regression.MPJ; Worksheet: Worksheet 1; 2/17/2020

Step	Instructions	Illustration
8	Prediction and confidence intervals • When using the equation, what would we expect the new prediction to be? The average prediction? • We can determine this by plotting prediction intervals and confidence levels. • Run the original problem again (stat/ regression/fitted line plot/options). • Select "options". • Select "display confidence interval" and "display prediction interval". • Select "ok" and "ok".	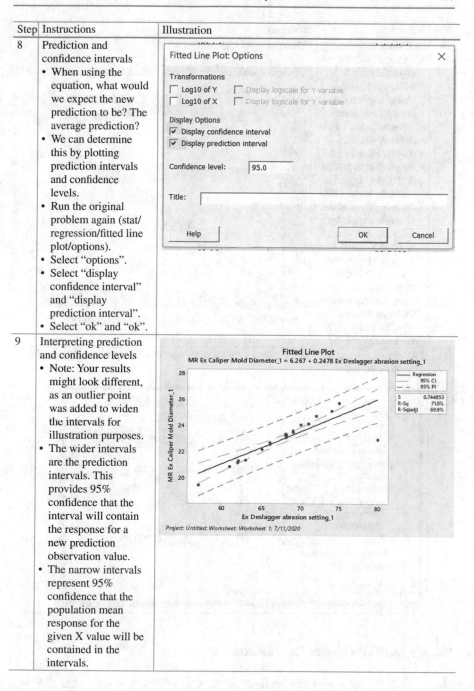
9	Interpreting prediction and confidence levels • Note: Your results might look different, as an outlier point was added to widen the intervals for illustration purposes. • The wider intervals are the prediction intervals. This provides 95% confidence that the interval will contain the response for a new prediction observation value. • The narrow intervals represent 95% confidence that the population mean response for the given X value will be contained in the intervals.	

In the illustration for step 8:

Fitted Line Plot: Options

Transformations
- ☐ Log10 of Y   ☐ Display logscale for Y variable
- ☐ Log10 of X   ☐ Display logscale for X variable

Display Options
- ☑ Display confidence interval
- ☑ Display prediction interval

Confidence level: 95.0

Title:

Help   OK   Cancel

In the illustration for step 9:

Fitted Line Plot
MR Ex Caliper Mold Diameter_1 = 6.267 + 0.2478 Ex Deslagger abrasion setting_1

Legend:
- Regression
- 95% CI
- 95% PI

S   0.744853
R-Sq   71.0%
R-Sq(adj) 69.9%

X-axis: Ex Deslagger abrasion setting_1
Y-axis: MR Ex Caliper Mold Diameter_1

Project: Untitled: Worksheet: Worksheet 1: 7/11/2020

Step	Instructions	Illustration
10	Prediction equation • Example: What is the expected caliper torsion rod Mold diameter when the Deslagger abrasion force setting is 70? • Answer: −0.0727 + 0.3435(70) = 23.97. • Remember to predict within the range. For example, instead of 79 (which was in the X source data range), we wanted to predict Y when X is 90? See the graph below the equation. It is not certain whether the trend will continue, increase, or decrease after the last point in the source data.	The regression equation: The regression equation is Ex Caliper Mold Diameter = - 0.0727 + 0.3435 Ex Deslagger abrasion setting Only predict in the data range: 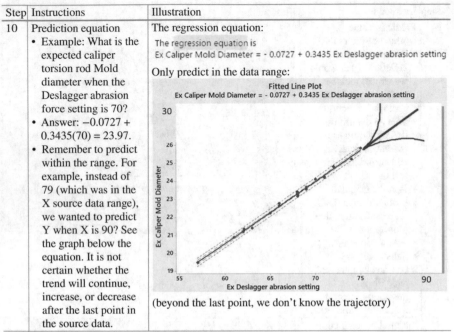 (beyond the last point, we don't know the trajectory)
11	*The project storyboard*: The following is an example how this information might appear in a project storyboard:

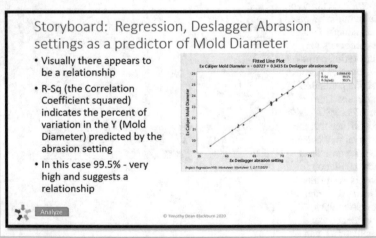

5.2.4   Multiple Linear Regression

But what if we have multiple predictors? Instead of $Y = a + bX$, what if $Y = a + b_1X_1 + b_2X_2 \ldots + b_nX_n$? In this case, we need to run a multiple regression. To illustrate, let's consider and expand upon the prior example for single linear regression, where the slag removal from the brake caliper mold was shown to be related to other deslagger settings.

After further discussions, Albert determined there might be other factors or predictors that should be considered. For example, some on the team believed ambient

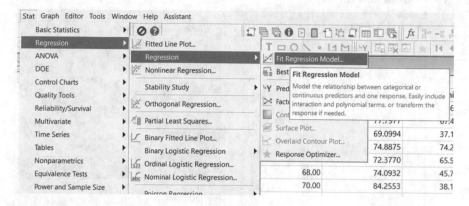

Fig. 5.21 Minitab path for multiple linear regression

temperature, relative humidity, and the lubricant pH could have an impact. Also, there were two deslagger machines used, and another team member felt they might be performing differently. So, the additional factors were added to the model for analysis.

To run a linear regression with multiple factors (this will also work with one predictor), go to the Minitab path Stat/Regression/Regression/Fit Regression Model. This is also shown in Fig. 5.21. Then the detailed step-by-step instructions are provided, along with key assumptions and considerations when performing multiple linear regression analysis.

Step	Instructions	Illustration
1	The data • Paste the multiple linear regression (from the regression tab in the dataset file) into Minitab.	
2	Performing the analysis • Minitab path stat/regression/regression/fit regression model. • In "response," select "MR ex caliper Mold Diameter_1". • In "continuous predictors," choose the deslagger abrasion setting, ambient temperature, relative humidity, and Lub pH. • In "categorical predictors," choose "Abraizer machine". • Select "graphs".	

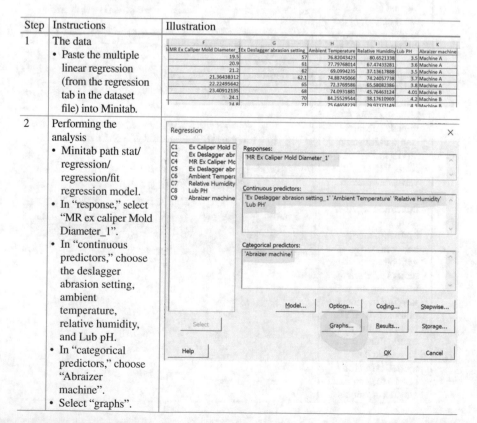

Step	Instructions	Illustration
3	• Select "four in one". • Select "ok" and "ok".	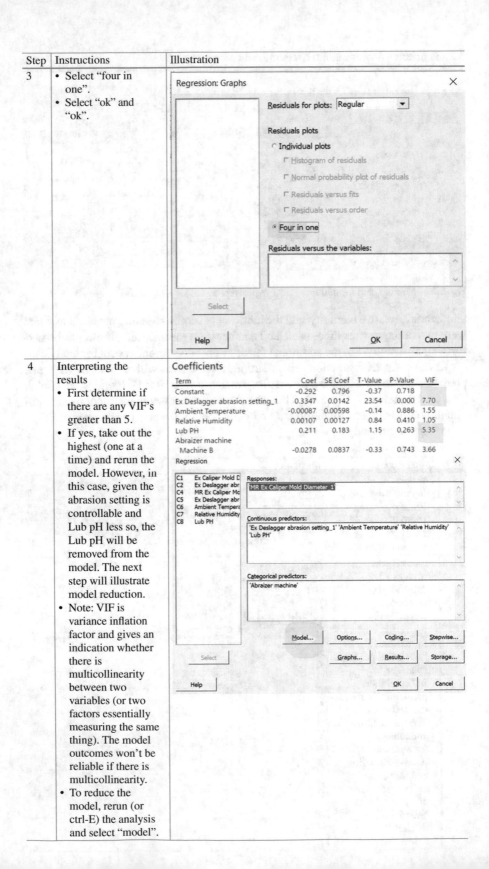
4	Interpreting the results • First determine if there are any VIF's greater than 5. • If yes, take out the highest (one at a time) and rerun the model. However, in this case, given the abrasion setting is controllable and Lub pH less so, the Lub pH will be removed from the model. The next step will illustrate model reduction. • Note: VIF is variance inflation factor and gives an indication whether there is multicollinearity between two variables (or two factors essentially measuring the same thing). The model outcomes won't be reliable if there is multicollinearity. • To reduce the model, rerun (or ctrl-E) the analysis and select "model".	**Coefficients** | Term	Coef	SE Coef	T-Value	P-Value	VIF
---	---	---	---	---	---
Constant	-0.292	0.796	-0.37	0.718	
Ex Deslagger abrasion setting_1	0.3347	0.0142	23.54	0.000	7.70
Ambient Temperature	-0.00087	0.00598	-0.14	0.886	1.55
Relative Humidity	0.00107	0.00127	0.84	0.410	1.05
Lub PH	0.211	0.183	1.15	0.263	5.35
Abraizer machine					
Machine B	-0.0278	0.0837	-0.33	0.743	3.66	

Step	Instructions	Illustration
5	Model reduction • Select "Lub pH". • Click the X to remove it form the next analysis. • Select "OK" and "OK".	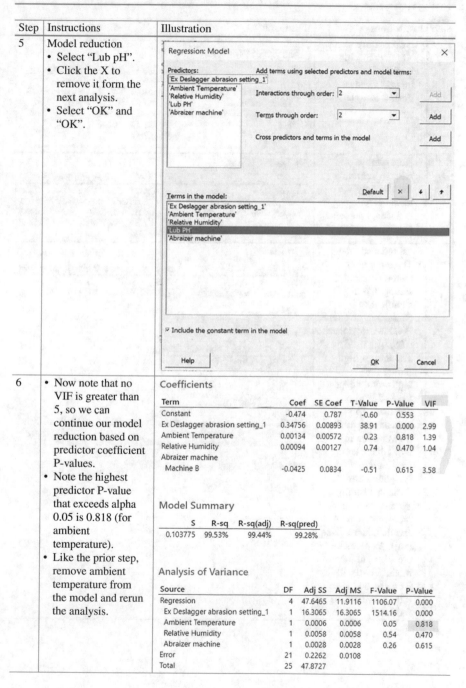
6	• Now note that no VIF is greater than 5, so we can continue our model reduction based on predictor coefficient P-values. • Note the highest predictor P-value that exceeds alpha 0.05 is 0.818 (for ambient temperature). • Like the prior step, remove ambient temperature from the model and rerun the analysis.	**Coefficients** (table below) **Model Summary** (table below) **Analysis of Variance** (table below)

Coefficients

Term	Coef	SE Coef	T-Value	P-Value	VIF
Constant	-0.474	0.787	-0.60	0.553	
Ex Deslagger abrasion setting_1	0.34756	0.00893	38.91	0.000	2.99
Ambient Temperature	0.00134	0.00572	0.23	0.818	1.39
Relative Humidity	0.00094	0.00127	0.74	0.470	1.04
Abraizer machine					
Machine B	-0.0425	0.0834	-0.51	0.615	3.58

Model Summary

S	R-sq	R-sq(adj)	R-sq(pred)
0.103775	99.53%	99.44%	99.28%

Analysis of Variance

Source	DF	Adj SS	Adj MS	F-Value	P-Value
Regression	4	47.6465	11.9116	1106.07	0.000
Ex Deslagger abrasion setting_1	1	16.3065	16.3065	1514.16	0.000
Ambient Temperature	1	0.0006	0.0006	0.05	0.818
Relative Humidity	1	0.0058	0.0058	0.54	0.470
Abraizer machine	1	0.0028	0.0028	0.26	0.615
Error	21	0.2262	0.0108		
Total	25	47.8727			

Step	Instructions	Illustration
7	• Next remove Abraizer machine, which has the highest P-value greater than 0.05.	**Analysis of Variance**
8	• Next remove relative humidity. • Interestingly, the model is reducing to a single predictor in this example.	**Analysis of Variance**
9	Final reduced model • The remaining factor, abrasion setting, has a significant coefficient given the P-value is less than 0.05. • The R-sq is high (unusually high—don't expect to see this in practice). This indicates 99.49% of the variation in mold diameter is predicted by variation in the deslagger setting. • When two or more predictors remain significant, use R-sq (adj). As predictors are added to the model, R-sq will increase, and R-sq (adj) compensates for that.	**Coefficients**

Step 7 Analysis of Variance:

Source	DF	Adj SS	Adj MS	F-Value	P-Value
Regression	3	47.6459	15.8820	1540.96	0.000
Ex Deslagger abrasion setting_1	1	17.1447	17.1447	1663.48	0.000
Relative Humidity	1	0.0054	0.0054	0.52	0.477
Abraizer machine	1	0.0022	0.0022	0.21	0.648
Error	22	0.2267	0.0103		
Total	25	47.8727			

Step 8 Analysis of Variance:

Source	DF	Adj SS	Adj MS	F-Value	P-Value
Regression	2	47.6437	23.8219	2393.04	0.000
Ex Deslagger abrasion setting_1	1	47.1309	47.1309	4734.57	0.000
Relative Humidity	1	0.0056	0.0056	0.56	0.463
Error	23	0.2290	0.0100		
Total	25	47.8727			

Step 9 Coefficients:

Term	Coef	SE Coef	T-Value	P-Value	VIF
Constant	-0.073	0.333	-0.22	0.829	
Ex Deslagger abrasion setting_1	0.34350	0.00492	69.82	0.000	1.00

Model Summary

S	R-sq	R-sq(adj)	R-sq(pred)
0.0988490	99.51%	99.49%	99.42%

Analysis of Variance

Source	DF	Adj SS	Adj MS	F-Value	P-Value
Regression	1	47.6382	47.6382	4875.41	0.000
Ex Deslagger abrasion setting_1	1	47.6382	47.6382	4875.41	0.000
Error	24	0.2345	0.0098		
Total	25	47.8727			

Step	Instructions	Illustration
10	Check assumptions (see the graph below)	

- Assumptions must be met to consider the regression valid.
- Normal probability plot: Residuals must be normally distributed (residuals are the differences between the actual and predicted values). P-value >0.05, so meets assumption.
- Versus fits: The predicted (fitted) values must show a random pattern versus residuals: Meets assumption.
- Versus order: The residuals must show a random pattern in the order of observation: Meets assumption.

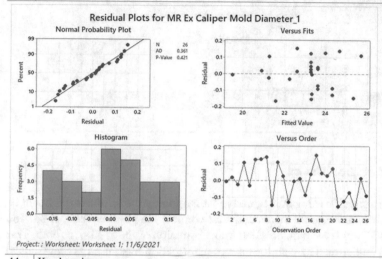

Residual Plots for MR Ex Caliper Mold Diameter_1

Project: ; Worksheet: Worksheet 1; 11/6/2021

11	Key learnings

- The Abraizer machine is not significant—Issues with the mold diameter are occurring when either are used.
- Environmental conditions are not predictive.
- But what about lubricant pH? This might be worth pursuing more. After reviewing this with a chemist, it was determined that in some instances heat can affect pH of this type of lubricant.
- That is, the stronger the abrasion force, the more the heat and the higher the pH. This is confirmed although there is a concern with non-normal residuals.
- However, this is really measuring the same thing—Abrasion setting, and the effect of the lubricant was ruled out.

12	Dealing with non-normal residuals

- In this example, the assumption for normally distributed residuals was met, but sometimes that is not the case.
- See the 4-in-1 plot below to illustrate a situation where the residuals are not normal. Imagine if we had seen this relationship between Lub pH and Deslagger setting (different than the prior example but shown to illustrate a point).
- Should this occur.
  - First confirm the point is valid. If so, you must leave it in.
  - If you leave it in, do a data transformation of the Y, X, or both. This requires trial and error. Path: Stat/control charts/Box-cox transformation.
- In this case the point was determined to be an error in data entry and was removed.

Step	Instructions	Illustration

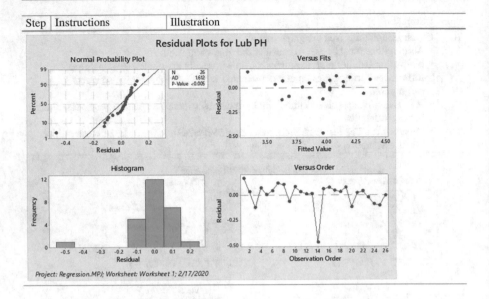

5.2.5   Regression Issues

Sometimes assumptions are not met, and most frequently the residuals might not be normally distributed. Should this occur, first confirm the point is valid. If so, you must leave it. If you leave it in, do a data transformation of the Y, X(s), or both. This requires trial and error. The path for data transformation most used is Stat/Control Charts/Box Cox Transformation.

Also, sometimes Minitab flags unusual points. While not in the prior examples, imagine when running regression for the abrasion setting as a predictor of mold diameter, and Minitab showed the "Fits and Diagnostics for Unusual Observations" as shown in Fig. 5.22. Also see Fig. 5.23 which shows a point that is influencing the regression line (or rotating it clockwise). This is an example of a point that overly influences the regression line.

Fits and Diagnostics for Unusual Observations

Obs	Ex Caliper Mold Diameter	Fit	Resid	Std Resid		
1	19.500	20.293	-0.793	-1.39		X
27	23.500	26.243	-2.743	-4.95	R	X

R Large residual
X Unusual X

Fig. 5.22 Unusual points example

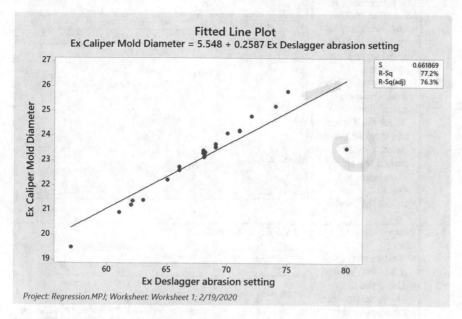

Fig. 5.23 Example of point influencing regression fit

Then referring to the output from Minitab in Fig. 5.22, note the flags include an "X" and "R." Minitab will flag observations that are unusual and have a large residual. "X" is indicated for points that may overly influence the regression line. "R" is indicated for large, standardized residuals greater than 2 standard deviations. In practice, review all "X" flags to confirm whether the point is accurate. Also evaluate any data points with "R" flags greater than 3 standard deviations.

Finally, a lack of sufficient data can be problematic. The amount of data needed depends on the variation in the data, but as a rule of thumb, linear regression has at least ten samples per predictor. Also most importantly, ensure the data is representative of the population and includes the full range you need to predict.

5.2.6   Correlation and Visualization in Minitab

While regression is considered more robust, correlation can also help us gain valuable insights in the data. The R-Sq in regression is the correlation coefficient squared. The following shows step-by-step instructions on how to perform correlation analysis and creates a helpful graph to further visualize correlation or relationships between variables.

Step	Instructions	Illustration
1	Performing the correlation analysis • Use the data from the prior multiple linear regression example. • Minitab path: Stat/ basic statistics/ correlation. • Select the X's first and then the Y last. • Select "results" and ensure all three boxes are checked. • Select "ok" and "ok".	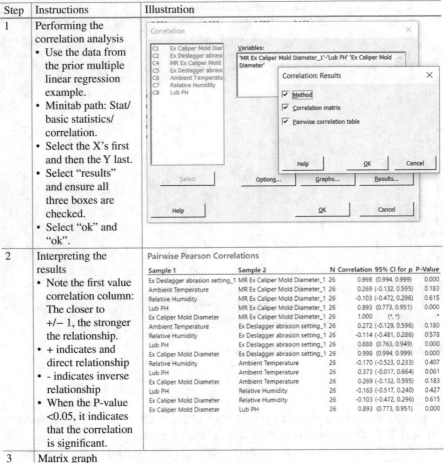

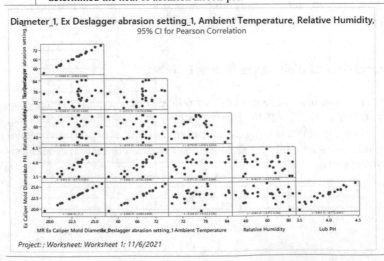

Step 2

Step	Instructions	Illustration
2	Interpreting the results • Note the first value correlation column: The closer to +/− 1, the stronger the relationship. • + indicates and direct relationship • - indicates inverse relationship • When the P-value <0.05, it indicates that the correlation is significant.	**Pairwise Pearson Correlations**

Pairwise Pearson Correlations

Sample 1	Sample 2	N	Correlation	95% CI for ρ	P-Value
Ex Deslagger abrasion setting_1	MR Ex Caliper Mold Diameter_1	26	0.998	(0.994, 0.999)	0.000
Ambient Temperature	MR Ex Caliper Mold Diameter_1	26	0.269	(-0.132, 0.595)	0.183
Relative Humidity	MR Ex Caliper Mold Diameter_1	26	-0.103	(-0.472, 0.296)	0.615
Lub PH	MR Ex Caliper Mold Diameter_1	26	0.893	(0.773, 0.951)	0.000
Ex Caliper Mold Diameter	MR Ex Caliper Mold Diameter_1	26	1.000	(*, *)	*
Ambient Temperature	Ex Deslagger abrasion setting_1	26	0.272	(-0.129, 0.596)	0.180
Relative Humidity	Ex Deslagger abrasion setting_1	26	-0.114	(-0.481, 0.286)	0.578
Lub PH	Ex Deslagger abrasion setting_1	26	0.888	(0.763, 0.949)	0.000
Ex Caliper Mold Diameter	Ex Deslagger abrasion setting_1	26	0.998	(0.994, 0.999)	0.000
Relative Humidity	Ambient Temperature	26	-0.170	(-0.523, 0.233)	0.407
Lub PH	Ambient Temperature	26	0.373	(-0.017, 0.664)	0.061
Ex Caliper Mold Diameter	Ambient Temperature	26	0.269	(-0.132, 0.595)	0.183
Lub PH	Relative Humidity	26	-0.163	(-0.517, 0.240)	0.427
Ex Caliper Mold Diameter	Relative Humidity	26	-0.103	(-0.472, 0.296)	0.615
Ex Caliper Mold Diameter	Lub PH	26	0.893	(0.773, 0.951)	0.000

Step	Instructions
3	Matrix graph • Automatically generates in Minitab V20. • For Minitab V18 (or to manually generate), go to Minitab path: Graph/Matrix Plot. • Appears to be a relationship between lube pH and abrasion settings to caliper mold diameter. • Appears to be relationship between Lube pH and the abrasion setting—Later determined the heat of abrasion affects pH.

Diameter_1, Ex Deslagger abrasion setting_1, Ambient Temperature, Relative Humidity,
95% CI for Pearson Correlation

Project: ; Worksheet: Worksheet 1; 11/6/2021

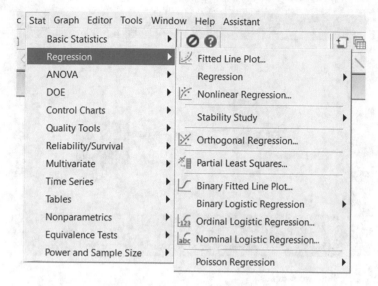

Fig. 5.24 Other regression techniques

5.2.7   Other Regression Tools (Introduction)

While beyond the scope of this book, there are other useful regression tools available in Minitab as shown in Fig. 5.24. Some of these are further explained as follows:
- Nonlinear regression: Helpful to determine if a dataset is predicted by a known nonlinear equation.
- Partial least squares (PLS): Used when there are a lot of factors versus sample size and/or when there is multicollinearity that results in an un-useful model.
- Binary logistic regression (BLR): This is used to predict pass or fail, or 1 or 0, or other binary responses.
- Ordinal logistic regression (OLR): Use this when predicting ordinal (Likert) data (e.g., 1, 2, 3, 4).
- Nominal logistic regression: Use this when the data consists of categories with no natural order (e.g., red, yellow, green).

5.3   Two Sample T Test, Mann-Whitney: Minitab Methods and Analysis Detail

5.3.1   Two Sample T Test Overview

A Two Sample T Test compares the means of two samples. It determines whether there is sufficient (statistical) evidence that there is a difference in the means. This is practically useful to prove there is a difference (e.g., before/after), such as confirming improvements were efficacious.

To visualize this, see the boxplot chart in Fig. 5.25. The boxplot chart includes a box with whiskers. Each whisker represents 25% of the data, and the box represents

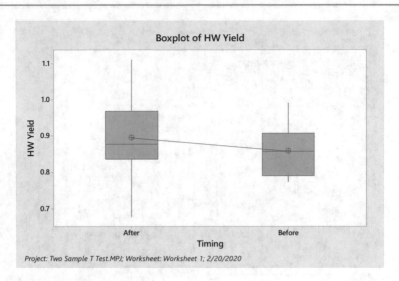

Fig. 5.25 Box plot, two samples

50% of the data, with the line in the box indicating the median. This is like turning a normal distribution curve on its side. (To generate a box plot in Minitab, go to Graphs/Boxplot.) This is an example that illustrates before and after yield. While the *after* seems to have a higher average or median yield, how can we be certain with statistical significance? Here the Two Sample T Test can be used to answer this question.

In a Two Sample T test, the null hypothesis, H_o asserts the means of the two sample sets are equal. The alternative hypothesis, or H_A, states the means of the two sample sets are not equal. To determine if we can reject the null hypothesis and conclude the means are different, we rely on the P-value. If the P-value is low (less than or equal to alpha = 0.05), there is 5% or less chance of rejecting the null incorrectly, so we reject it and accept the alternative hypothesis. That is, we conclude there is a difference in the means. However, if the P-value is not low, we cannot conclude the two sample sets are unequal. When this occurs, it could be they are essentially the same, or there is too much variation in the data to detect if they are unequal, or there is insufficient data to determine if they are equal. It *is incorrect to say, when the P-value is > 0.05, "The means are equal"*—instead say, there is insufficient statistical evidence to conclude the means are not equal. Speaking this way is precise and needed to explain statistical outcomes correctly but might not at first seem to be a natural way of speaking.

To illustrate how to apply this practically, let's do an example from KIND Karz. Remember the control chart in Measure indicated after the assembler was put in, the average defect rate increased. But is this statistically significant? A Two Sample T Test can help with this if the normality assumption is met.

Note: This data in this example is discrete data (discrete proportion), but given it behaves as normal data, a Two Sample T Test could be used. Go to Two Sample T Test dataset and paste into Minitab. The Minitab path for a Two Sample T Test is Stat/Basic Statistics/2-Sample t or as shown in Fig. 5.26.

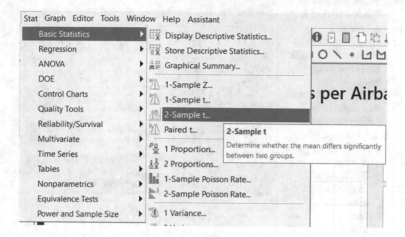

Fig. 5.26 Minitab path, Two Sample T Test

5.3.2 Two Sample T Test in Minitab

Step	Instructions	Illustration
4	Sample data • Paste the data into Minitab from the two sample T test dataset.	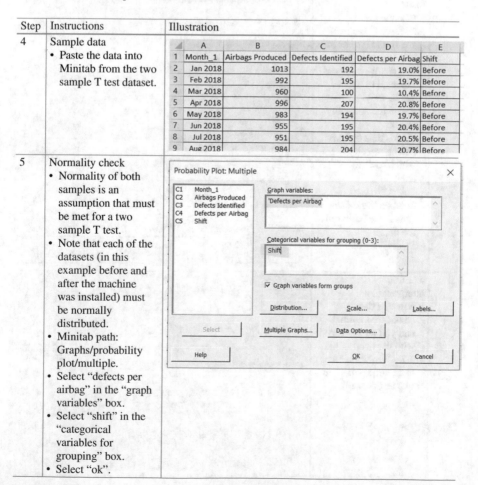
5	Normality check • Normality of both samples is an assumption that must be met for a two sample T test. • Note that each of the datasets (in this example before and after the machine was installed) must be normally distributed. • Minitab path: Graphs/probability plot/multiple. • Select "defects per airbag" in the "graph variables" box. • Select "shift" in the "categorical variables for grouping" box. • Select "ok".	

Step	Instructions	Illustration
6	Normality assumptions • The before data is not normal. • To be normal, most dots should be inside the confidence bands. • Both P-values should be > alpha 0.05. • Order of preference as what to do next: o See if outlier point is correct. o If yes, use a nonparametric test. o Data transformation is possible but should be used as the final option.	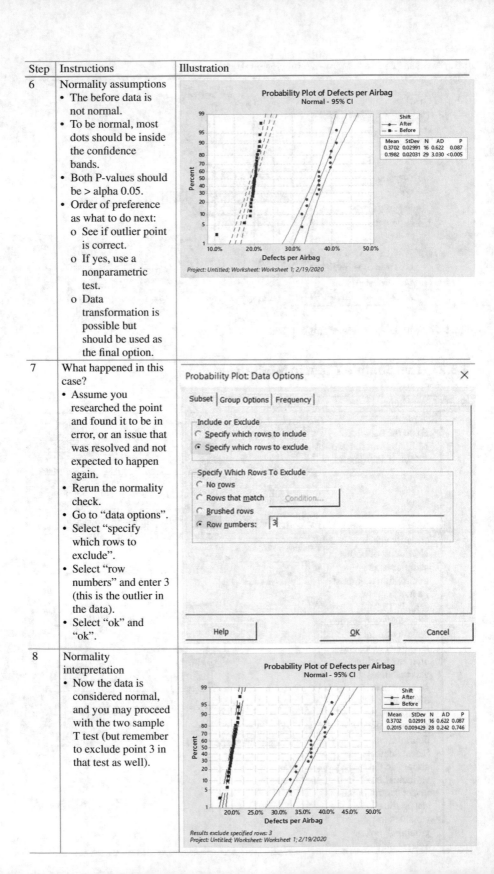
7	What happened in this case? • Assume you researched the point and found it to be in error, or an issue that was resolved and not expected to happen again. • Rerun the normality check. • Go to "data options". • Select "specify which rows to exclude". • Select "row numbers" and enter 3 (this is the outlier in the data). • Select "ok" and "ok".	
8	Normality interpretation • Now the data is considered normal, and you may proceed with the two sample T test (but remember to exclude point 3 in that test as well).	

Step	Instructions	Illustration
9	Two sample T test • Create a new worksheet in Minitab (file/new/worksheet), paste the data, but remove point 3. • Minitab path for two sample T test: Stat/ basic statistics/2-sample t. • In the top drop down, select "both samples are in one column". • In "samples," select "defects per airbag". • In sample IDs, select "shift". • Select "graphs".	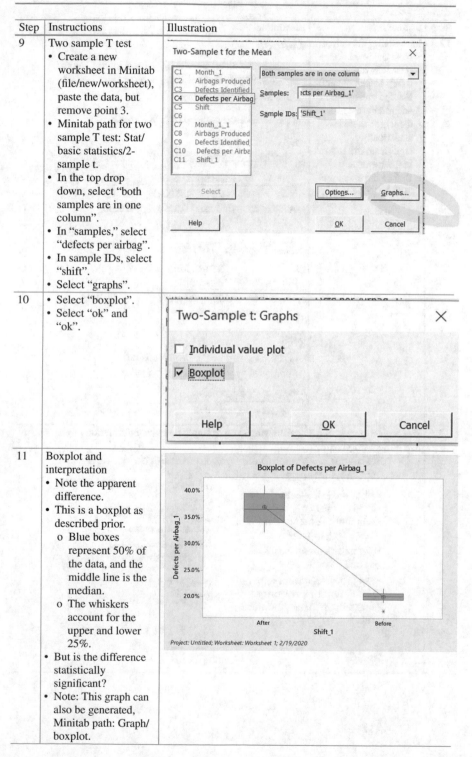
10	• Select "boxplot". • Select "ok" and "ok".	
11	Boxplot and interpretation • Note the apparent difference. • This is a boxplot as described prior. o Blue boxes represent 50% of the data, and the middle line is the median. o The whiskers account for the upper and lower 25%. • But is the difference statistically significant? • Note: This graph can also be generated, Minitab path: Graph/ boxplot.	

Step	Instructions	Illustration
12	Interpreting the output • Note the low P-value (less than alpha 0.05): Reject the null that the means before and after are equal. • Conclude there is a statistically significant difference in the means. • Before the mean was 20% but 37% now.	**Method** μ_1: mean of Defects per Airbag_1 when Shift_1 = After μ_2: mean of Defects per Airbag_1 when Shift_1 = Before Difference: $\mu_1 - \mu_2$ *Equal variances are not assumed for this analysis.* **Descriptive Statistics: Defects per Airbag_1** Shift_1 N Mean StDev SE Mean After 16 0.3702 0.0299 0.0075 Before 28 0.20152 0.00943 0.0018 **Estimation for Difference** 95% CI for Difference Difference 0.16865 (0.15235, 0.18494) **Test** Null hypothesis H_0: $\mu_1 - \mu_2 = 0$ Alternative hypothesis H_1: $\mu_1 - \mu_2 \neq 0$ T-Value DF P-Value 21.94 16 0.000
13	*The project storyboard*: The following is an example how this information might appear in a project storyboard: Storyboard: Two Sample T-Test, confirming defect rate increased after new assembler installed • Note the low P-value (less than alpha 0.05): reject the null that the means before and after are equal • Before the mean was 20%, abut 37% now • Confirms the new assembler is different that the prior one, and has a worst defect rate Two-Sample T-Test and CI: Defects per Airbag_1, Shift_1 Method μ_1: mean of Defects per Airbag_1 when Shift_1 = After μ_2: mean of Defects per Airbag_1 when Shift_1 = Before Difference: $\mu_1 - \mu_2$ *Equal variances are not assumed for this analysis.* Descriptive Statistics: Defects per Airbag_1 Shift_1 N Mean StDev SE Mean After 16 0.3702 0.0299 0.0075 Before 28 0.20152 0.00943 0.0018 Estimation for Difference 95% CI for Difference Difference 0.16865 (0.15235, 0.18494) Test Null hypothesis H_0: $\mu_1 - \mu_2 = 0$ Alternative hypothesis H_1: $\mu_1 - \mu_2 \neq 0$ T-Value DF P-Value 21.94 16 0.000 © Timothy Dean Blackburn 2020 33

5.3.3   Mann-Whitney Test in Minitab

Imagine if we could not remove the outlier point. What to do next? The next step would be to run a nonparametric alternative test to the Two Sample T Test. This is called the Mann-Whitney test, which evaluates to see if there is a statistically significant difference in the *medians* versus the means.

First, let's understand the concept of considering medians versus means. Imagine you had the following numbers {1, 3, 2, 2, 1, 3, 1, 1, 2, 1, 3, 1, 3, 3,100,000,000} and wanted to understand the central tendency of the dataset. The mean is 6,666,668, and the median is 2 which is much better description of the center of the data. The median then, where there are outliers, is a better representation of the central tendency of the data.

The Mann-Whitney test is a nonparametric alternative to the Two Sample T Test. (Nonparametric refers to it not being normal or fitting a typical distribution.) To run Mann-Whitney, go to Stat/Nonparametrics/Mann-Whitney, as shown in Fig. 5.27.

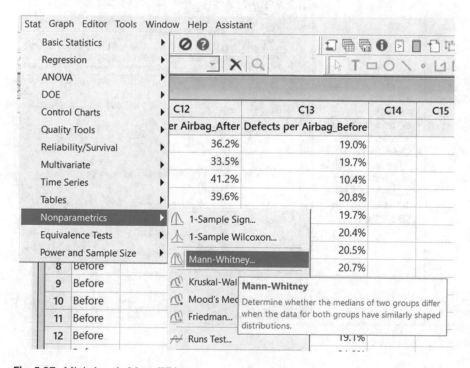

Fig. 5.27   Minitab path, Mann-Whitney

Step	Instructions	Illustration
1	Sample data • Paste the data into Minitab from the two sample T test dataset, or return to the worksheet with the original data. • This should be the original data with the outlier point left in.	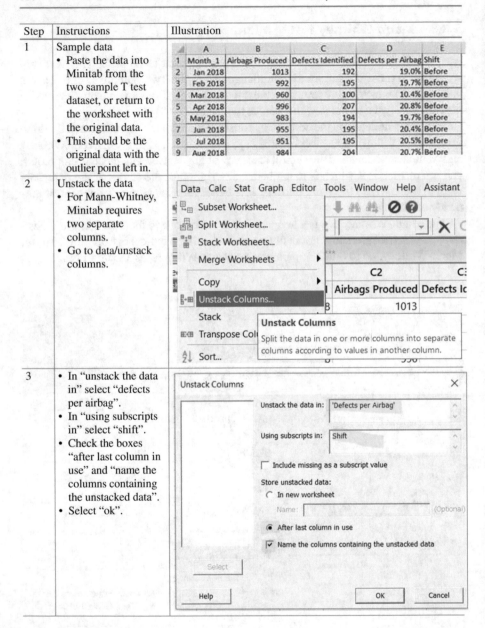
2	Unstack the data • For Mann-Whitney, Minitab requires two separate columns. • Go to data/unstack columns.	
3	• In "unstack the data in" select "defects per airbag". • In "using subscripts in" select "shift". • Check the boxes "after last column in use" and "name the columns containing the unstacked data". • Select "ok".	

Step	Instructions	Illustration
4	Run Mann-Whitney • Minitab path: Stat/ Nonparametrics/ Mann-Whitney. • Enter the two new sample columns in "first" and "second sample". • Retain 95% for "confidence level". • Select "not equal" for "alternative". • Select "ok".	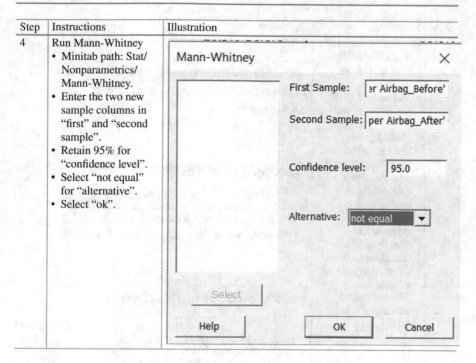

Step	Instructions	Illustration
5	Interpreting the results • Mann-Whitney medians (versus means). • This test is typically less powerful test than two sample T test (e.g., more likely to fail to reject the null when there really is a difference) but needed when the data is nonparametric. • Note P-value less than alpha 0.05, so we can conclude the medians are different (36.7% after and 20.1% before).	**Method** η_1: median of Defects per Airbag_Before η_2: median of Defects per Airbag_After Difference: $\eta_1 - \eta_2$ **Descriptive Statistics** <table><tr><th>Sample</th><th>N</th><th>Median</th></tr><tr><td>Defects per Airbag_Before</td><td>29</td><td>0.201369</td></tr><tr><td>Defects per Airbag_After</td><td>16</td><td>0.366876</td></tr></table> **Estimation for Difference** <table><tr><th>Difference</th><th>CI for Difference</th><th>Achieved Confidence</th></tr><tr><td>-0.169521</td><td>(-0.191693, -0.158477)</td><td>95.23%</td></tr></table> **Test** Null hypothesis H_0: $\eta_1 - \eta_2 = 0$ Alternative hypothesis H_1: $\eta_1 - \eta_2 \neq 0$ <table><tr><th>Method</th><th>W-Value</th><th>P-Value</th></tr><tr><td>Not adjusted for ties</td><td>435.00</td><td>0.000</td></tr><tr><td>Adjusted for ties</td><td>435.00</td><td>0.000</td></tr></table>

Step	Instructions	Illustration
6	Checking the similarly shaped distributions assumption • Mann-Whitney assumes the distributions are similarly shaped. • Minitab path: Graphs/histogram/ with groups. • In "graph variables," select "defects per airbag". • In "categorical variables for grouping," select "shift". • Select "multiple graphs".	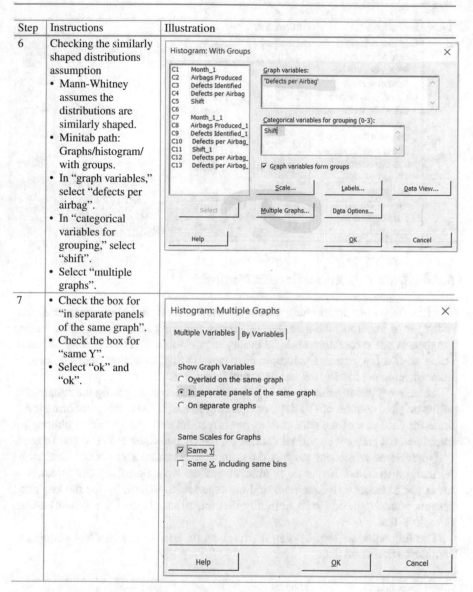
7	• Check the box for "in separate panels of the same graph". • Check the box for "same Y". • Select "ok" and "ok".	

Step	Instructions	Illustration
8	• In this case, it is hard to tell but seems both to be right centered. • Usually, a concern would be when the shapes are clearly different (e.g., when one is right centered and the other left). • Therefore, the assumptions of similar shaped distributions are accepted.	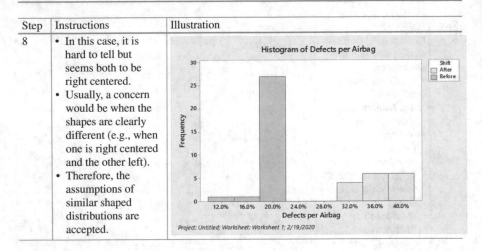

5.3.4   Data Transformations in Minitab

The last resort after determining a data is normal (after trying a nonparametric alternative) is to transform the data. But again, try the other steps before transforming the data. While it sometimes it is necessary to perform a data transformation, avoid doing so if a few extreme values, excessive rounding, or values below a level of quantification are forcing the use of a transformation.

There are a variety of methods to transform data, such as getting the square root of the original column of (Y) data, or squaring it, etc. If the same transformation is done for each row of the data, it does not affect the underlying dataset although it might make it perform as normal data. If so, then a Two Sample T Test can be used.

Note: Some references suggest datasets over a certain size do not need to be normally distributed. However, significant outliers from normality can affect even moderate or larger size datasets based on experience. Especially for the beginner, assume your data needs to be normally distributed to rely on the results of a Two Sample T Test.

The following are step-by-step instructions on how to use two key features in Minitab to transform data.

Step	Instructions	Illustration				
1	Sample data • Paste the data into Minitab from the two sample T test dataset, or return to the worksheet with the original data. This should be the original data with the outlier point left in		A	B	C	D	E
		1	Month_1	Airbags Produced	Defects Identified	Defects per Airbag	Shift
		2	Jan 2018	1013	192	19.0%	Before
		3	Feb 2018	992	195	19.7%	Before
		4	Mar 2018	960	100	10.4%	Before
		5	Apr 2018	996	207	20.8%	Before
		6	May 2018	983	194	19.7%	Before
		7	Jun 2018	955	195	20.4%	Before
		8	Jul 2018	951	195	20.5%	Before
		9	Aug 2018	984	204	20.7%	Before

Step	Instructions	Illustration
2	Box-Cox transformation • Create a new heading in the Minitab worksheet, "BoxCox trans defect rate". • Note: Box-cox requires original data values to be greater than zero. If zero or negative, go to the "Johnson's transformation," or do a manual transformation (e.g., add values to get above 1 and then transform).	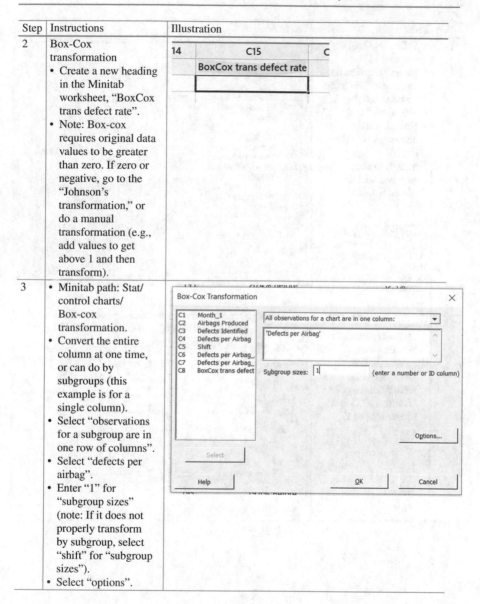
3	• Minitab path: Stat/ control charts/ Box-cox transformation. • Convert the entire column at one time, or can do by subgroups (this example is for a single column). • Select "observations for a subgroup are in one row of columns". • Select "defects per airbag". • Enter "1" for "subgroup sizes" (note: If it does not properly transform by subgroup, select "shift" for "subgroup sizes"). • Select "options".	

Step	Instructions	Illustration
4	• Select "optimal or rounded l". • Select "BoxCox trans defect rate" in "store transformed data in". • Select "ok" and "ok". • Note: Lambda corresponds to the type of transformation Box-cox will do as shown below. 	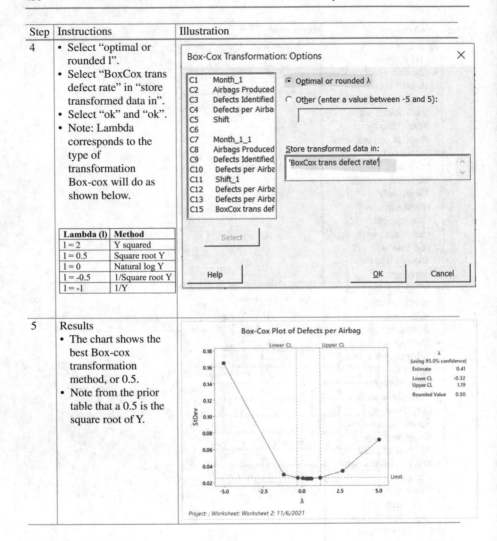
5	Results • The chart shows the best Box-cox transformation method, or 0.5. • Note from the prior table that a 0.5 is the square root of Y.	

Lambda (l)	Method
l = 2	Y squared
l = 0.5	Square root Y
l = 0	Natural log Y
l = -0.5	1/Square root Y
l = -1	1/Y

Step	Instructions	Illustration
6	• Note the transformed data are stored in Minitab's worksheet view. • erun normality test and notice the transformed data is normally distributed. • Now run the two sample T test.	Probability Plot of BoxCox trans defect rate Normal - 95% CI Mean StDev N AD P 0.6079 0.02472 16 0.633 0.081 0.4488 0.01059 28 0.267 0.662 Results exclude specified rows: 3 Project: : Worksheet: Worksheet 2: 11/6/2021
7	Two Sample T test results • See prior steps for details. Outcomes are shown here. • P-value is < alpha 0.05 so conclude there is a statistically significant difference before and after the machine was installed. • However, the means don't make sense practically given they are transformed data. These would need to be untransformed (e.g., for this example, square the means). • But what if the Box cox wouldn't transform? • Then try Johnsons transformation.	**Method** μ_1: population mean of BoxCox trans defect rate when Shift = After μ_2: population mean of BoxCox trans defect rate when Shift = Before Difference: $\mu_1 - \mu_2$ *Equal variances are not assumed for this analysis.* **Descriptive Statistics: BoxCox trans defect rate** Shift N Mean StDev SE Mean After 16 0.6079 0.0247 0.0062 Before 29 0.4444 0.0256 0.0048 **Estimation for Difference** 95% CI for Difference Difference 0.16350 (0.14761, 0.17938) **Test** Null hypothesis $H_0: \mu_1 - \mu_2 = 0$ Alternative hypothesis $H_1: \mu_1 - \mu_2 \neq 0$ T-Value DF P-Value 20.97 32 0.000

Step	Instructions	Illustration
8	Johnson's transformation • If Box-Cox is not successful, then try Johnson's transformation. However, Box-cox is usually preferred as the *un-transformation* is simpler, given Johnson's uses a complex formula. • Create a new column header for the transformed data to be stored in the Minitab worksheet. • Minitab path: Stat/ quality tools/ Johnson's transformation. • Select "single column" and "defects per airbag". • Select the new column header for "store transformed data in". • Select "ok".	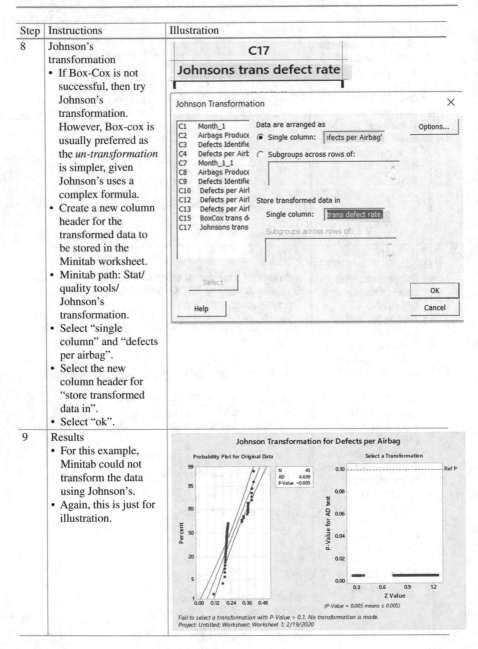
9	Results • For this example, Minitab could not transform the data using Johnson's. • Again, this is just for illustration.	

5.3.5   Sample Size Determination in Minitab

The sample size needed for a Two Sample T Test depends on the data variation, the desired level of detection, and the probability of detection when there really is a difference (typically 80%). Use the sample size feature in Minitab to determine the recommended sample size (or Minitab path: Stat/Power and Sample Size/2 Sample t) and as shown in Fig. 5.28. A step-by-step example is provided below.

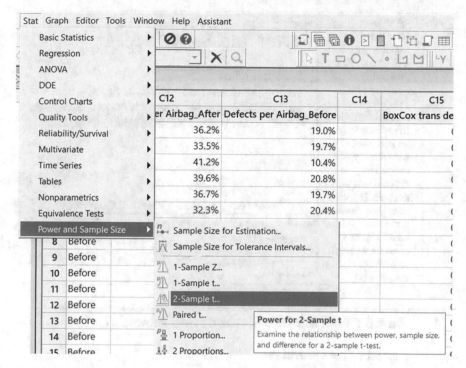

Fig. 5.28 Two Sample T Test power and sample size

Step	Instructions	Illustration
1	• Minitab path: Stat/ power and sample size/2 sample t. • Enter the minimum difference you would like to be able to detect. For this example, let's assume we would like to be able to detect 1% defect rate differences or 0.01. • Enter the power value. 0.8 is common. This will be explained later. • Determine the standard deviation using historical data. The Minitab path that can be used is stat/basic statistics/display descriptive statistics.	**Power and Sample Size for 2-Sample t** ✕ Specify values for any two of the following: Sample sizes: [ ] Differences: [.01] Power values: [.8] Standard deviation: [.00943] Options... Graph... Help OK Cancel

Step	Instructions	Illustration
2	• Note here 15 samples per group will give an 80% probability of detecting at least 1% difference in mean defect rates. • In our case it's a bit irrelevant as we have already showed a difference in prior analysis. But what if had not shown a difference? This could give us a clue that we need to get more samples, etc.	**Power and Sample Size** 2-Sample t Test Testing mean 1 = mean 2 (versus ≠) Calculating power for mean 1 = mean 2 + difference α = 0.05 Assumed standard deviation = 0.00943 **Results** $\begin{array}{cccc} & \text{Sample} & \text{Target} & \\ \text{Difference} & \text{Size} & \text{Power} & \text{Actual Power} \\ 0.01 & 15 & 0.8 & 0.800477 \end{array}$ *The sample size is for each group.*

5.4   Paired T Test: Minitab Methods and Analysis Detail

5.4.1   Paired T Test Overview

An assumption of the Two Sample T Test is the two datasets are independent. That is, the rows for the two columns of data are not related in the sense one is predictive of the other. There isn't a natural paired order.

But what if they are not independent? For example, let's consider a situation from the KIND Karz case study. KIND Karz has been experiencing unusually higher failure rates of airbags. It was discovered that fabric thickness is more variable (and thinner) for airbags that failed than those that did not. There are two devices used to calibrate the equipment to ensure the fabric is produced with the correct thickness. It was discovered that Device B was used to calibrate fabric thickness settings for equipment making airbags that failed and Device A was used on those that had no failures. Is there a difference in the devices? To determine this, Albert decided to test the same swaths of airbag fabric on each device. This is paired data, and each data pair is dependent—that is, each swath will be measured using two different devices, with the intent of determining if the devices provide different readings. In this case, there will also be two columns of data, but each row will be measuring the unique fabric swath. The analysis method will be a paired T Test.

Unlike a Two Sample T Test, responses for a paired T Test are not independent but dependent. For each sample row, there is something common. In the example that will follow, Device A and B will measure same samples of airbag fabric (see Fig. 5.29). Also note there is a column of differences in readings (these differences must be normally distributed, not the original data).

Sample	Device A	Device B	Differences
1	0.0308371	0.0339208	0.0030837
2	0.0288898	0.0317788	0.0028890
3	0.0301070	0.0331177	0.0030107
4	0.0293784	0.0323162	0.0029378
5	0.0295244	0.0324769	0.0029524
6	0.0310079	0.0341087	0.0031008
7	0.0312110	0.0343321	0.0031211
8	0.0293853	0.0323239	0.0029385
9	0.0304669	0.0335135	0.0030467
10	0.0316544	0.0348198	0.0031654

Fig. 5.29 Paired T Test example data

For a paired T Test, the null hypothesis (H_o) is as follows: There is no differences in the paired groups (the average paircd differences are 0). Then the alternative hypothesis (H_A) is: There are differences in the paired groups (the average paired differences are not 0). If the P-value is low (less than or equal to alpha 0.05), there is 5% or less chance of rejecting the null incorrectly and conclude there is a difference. However, if the P-value is not low (≥ 0.05), you cannot conclude the two sample sets are unequal. Possibilities include the following when we fail to reject the null hypothesis: they are essentially equal, or there is too much variation in the data to detect if they are unequal, or there is insufficient data to determine if they are equal. *So, it is incorrect to say, when the P-value is > 0.05, "The means are equal."*

5.4.2 Paired T Test in Minitab

The Minitab path for a paired T Test is Stat/Basic Statistics/Paired t (or as shown in Fig. 5.30). Below are step-by-step instructions.

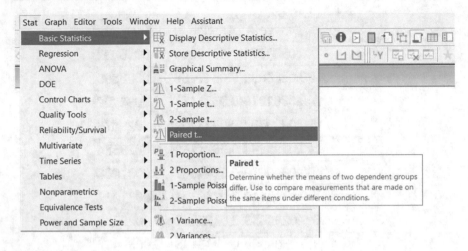

Fig. 5.30 Minitab path for paired T Test

Step	Instructions	Illustration
1	• Paste the data into Minitab from the paired T test dataset. • Note the differences are already calculated here. For future reference, calculate these in excel before pasting, or use the Calc/ calculator feature in Minitab.	<table><tr><th>Sample</th><th>Device A</th><th>Device B</th><th>Differences</th></tr><tr><td>1</td><td>0.0308371</td><td>0.0339208</td><td>0.0030837</td></tr><tr><td>2</td><td>0.0288898</td><td>0.0317788</td><td>0.0028890</td></tr><tr><td>3</td><td>0.0301070</td><td>0.0331177</td><td>0.0030107</td></tr><tr><td>4</td><td>0.0293784</td><td>0.0323162</td><td>0.0029378</td></tr><tr><td>5</td><td>0.0295244</td><td>0.0324769</td><td>0.0029524</td></tr><tr><td>6</td><td>0.0310079</td><td>0.0341087</td><td>0.0031008</td></tr><tr><td>7</td><td>0.0312110</td><td>0.0343321</td><td>0.0031211</td></tr><tr><td>8</td><td>0.0293853</td><td>0.0323239</td><td>0.0029385</td></tr><tr><td>9</td><td>0.0304669</td><td>0.0335135</td><td>0.0030467</td></tr><tr><td>10</td><td>0.0316544</td><td>0.0348198</td><td>0.0031654</td></tr></table>
2	Check if the differences are normally distributed • As noted prior, the differences must be normally distributed, not the original data. • Minitab path: Graph/ probability plot. • All points are within the confidence bands, and the P-value is > alpha = 0.05: Accept that the normality assumption is met.	**Probability Plot of Differences** Normal - 95% CI Mean 0.003025 StDev 0.00009289 N 10 AD 0.273 P-Value 0.582 *Percent* vs *Differences* (0.0027 to 0.0034) Project: Untitled; Worksheet: Worksheet 1; 2/20/2020

Step	Instructions	Illustration
3	Performing the analysis • Minitab path: Stat/ basic statistics/paired t. • Select "each sample is in a column". • Enter "device A" for "sample 1". • Enter "device B" for "sample 2". • Select "graphs".	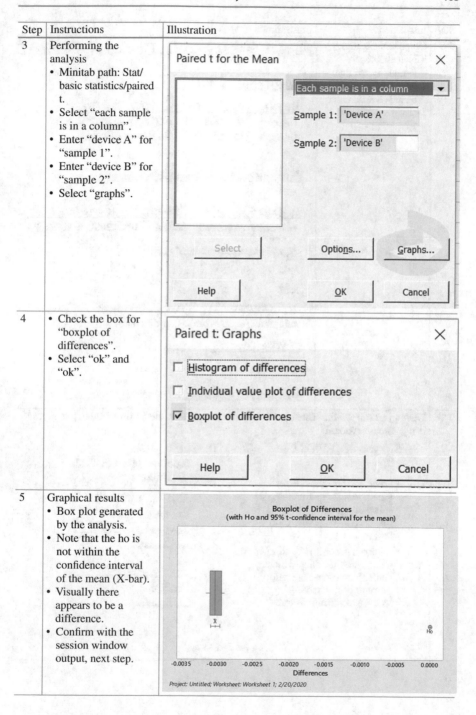
4	• Check the box for "boxplot of differences". • Select "ok" and "ok".	
5	Graphical results • Box plot generated by the analysis. • Note that the ho is not within the confidence interval of the mean (X-bar). • Visually there appears to be a difference. • Confirm with the session window output, next step.	

Step	Instructions	Illustration
6	Output conclusions • Note that the P-value is less than 0.05, so conclude there is a difference.	**Paired T-Test and CI: Device A, Device B** **Descriptive Statistics** Sample · N · Mean · StDev · SE Mean Device A · 10 · 0.030246 · 0.000929 · 0.000294 Device B · 10 · 0.033271 · 0.001022 · 0.000323 **Estimation for Paired Difference** Mean · StDev · SE Mean · 95% CI for μ_difference -0.003025 · 0.000093 · 0.000029 · (-0.003091, -0.002958) *μ_difference: mean of (Device A - Device B)* **Test** Null hypothesis $H_0: \mu\_difference = 0$ Alternative hypothesis $H_1: \mu\_difference \neq 0$ T-Value · P-Value -102.96 · 0.000 **Boxplot of Differences**
7	*The project storyboard*: The following is an example how this information might appear in a project storyboard:	

Storyboard: Paired T Test – Fabric Thickness Tester

Paired T-Test and CI: Device A, Device B

Descriptive Statistics

Sample   N   Mean   StDev   SE Mean
Device A   10   0.030246   0.000929   0.000294
Device B   10   0.033271   0.001022   0.000323

• Device B thickness tester (used on airbags that failed) was compared to Device A (used on airbags that didn't fail)

Estimation for Paired Difference

95% CI for
Mean   StDev   SE Mean   μ_difference
-0.003025   0.000093   0.000029   (-0.003091, -0.002958)

μ_difference: mean of (Device A - Device B)

• Device B is confirmed to give a false high reading, which leads to an over-calibration of the fabric machine, leading to thinner fabric

Test

Null hypothesis $H_0: \mu\_difference = 0$
Alternative hypothesis $H_1: \mu\_difference \neq 0$

T-Value   P-Value
-102.96   0.000

Boxplot of Differences

Analyze

10

5.4.3   One Sample Wilcoxon in Minitab (Nonparametric Alternative to a Paired T Test)

As noted prior, the differences must be normally distributed to run a paired T Test. If they are not normally distributed, first confirm the data is accurate. If so, then determine if the point(s) causing the non-normal behavior (where there was a special cause event) that has been resolved and is not expected to reoccur. If so, remove it and check again for normality. Otherwise, run a 1 Sample Wilcoxon, which is a nonparametric alternative. (If the 1 Sample Wilcoxon test cannot be run, then consider a data transformation as described for the Two Sample T Test section.)

The Minitab path for a 1 Sample Wilcoxon test is Stat/Nonparametrics/1-Sample Wilcoxon, or as shown in Fig. 5.31. Below are step-by-step instructions.

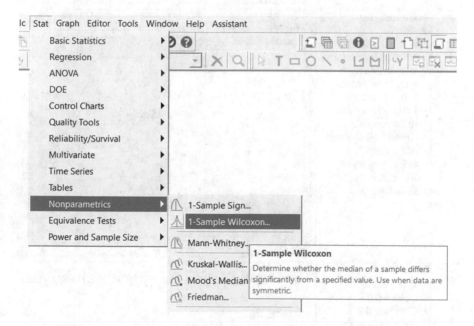

Fig. 5.31 Minitab path for 1 sample Wilcoxon test

Step	Instructions	Illustration
1	• Paste the data in Minitab's worksheet from the paired T test dataset. This is the same data that we used for the prior paired T test. • Note the differences are already calculated here. For future reference, calculate these in excel before pasting, or use the Calc/calculator feature in Minitab.	

Sample	Device A	Device B	Differences
1	0.0308371	0.0339208	0.0030837
2	0.0288898	0.0317788	0.0028890
3	0.0301070	0.0331177	0.0030107
4	0.0293784	0.0323162	0.0029378
5	0.0295244	0.0324769	0.0029524
6	0.0310079	0.0341087	0.0031008
7	0.0312110	0.0343321	0.0031211
8	0.0293853	0.0323239	0.0029385
9	0.0304669	0.0335135	0.0030467
10	0.0316544	0.0348198	0.0031654

Step	Instructions	Illustration
2	Performing the analysis • Minitab path: Stat/Nonparametrics/1-sample Wilcoxon select "each sample is in a column". • Enter "differences" in "variables". • Check "test median" and enter "0". • Select "not equal" for "alternative". • Select "ok".	**1-Sample Wilcoxon** ✕ Variables: Differences ○ Confidence interval Level: 95.0 ● Test median: 0.0 Alternative: not equal ▼ Select Help OK Cancel

Step	Instructions	Illustration
3	Results • P-value is < alpha = 0.05. • Conclude the differences in the paired values are not 0. • Therefore, conclude there is a difference.	

Wilcoxon Signed Rank Test: Differences

Method

η: median of Differences

Descriptive Statistics

Sample	N	Median
Differences	10	0.0030266

Test

Null hypothesis	$H_0: \eta = 0$
Alternative hypothesis	$H_1: \eta \neq 0$

Sample	N for Test	Wilcoxon Statistic	P-Value
Differences	10	55.00	0.006

Step	Instructions	Illustration
4	Confirm assumption of symmetry is met • A 1 sample Wilcoxon test requires that the differences be symmetrically distributed. • Minitab path: Graph/histogram. • Create a histogram of the differences. • Note the distribution is symmetrical (e.g., no apparent left or right shift), so the assumption is met, and the results of the 1 sample Wilcoxon test can be accepted.	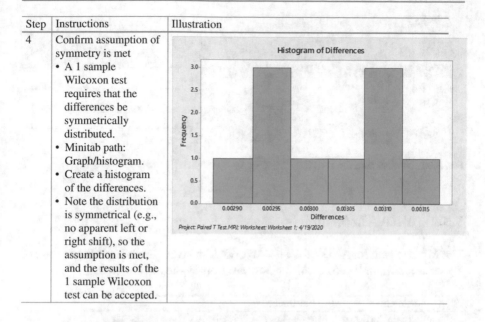

5.5   ANOVA, ANOM: Minitab Methods and Analysis Detail

5.5.1   ANOVA and ANOM Overview

In a prior section, we reviewed statistical tools to compare the means from two sample sets. But what do we do with more than two sample sets? If we compare two at a time, we increase the likelihood of rejecting falsely (the error compounds). Analysis of variance (ANOVA) is a technique that compares more than two sample means and determines whether one or more sample means is different than another mean. Analysis of means (ANOM) is similar but determines whether a mean is different than the mean of the means.

First, let's consider ANOVA. The null hypothesis (Ho) is: All averages are equal. The alternative hypothesis (H_A) is: At least one group average is different from another. If the P-value < alpha 0.05, reject the null and accept the alternative hypothesis.

The assumptions for ANOVA (and ANOM) are each group of data is normally distributed, and there are equal variances across groups of data. ANOVA can still be run if the variances are note equal, but it requires a change to a default.

To illustrate, consider an example from KIND Karz. The brake caliper torsion rod keeps breaking as evidenced by warrantee claims. One potential root cause is that the KIND Karz patented cast iron alloy is flawed, leading to a lower tensile strength and resulting in failures. Coupons of the material from retained samples will be compared. (Coupons are small samples of the actual material for testing.) Lot 1 include coupons from lots that failed, and Lots 2–4 are from lots that did not fail. Albert would like to compare mean tensile strength (Mpa) and see if there is a difference. (For good cast iron alloy of the type KIND Karz has patented, the average tensile strength is expected to be about 414 Mpa.) To analyze this, we will use the ANOVA tool in Minitab.

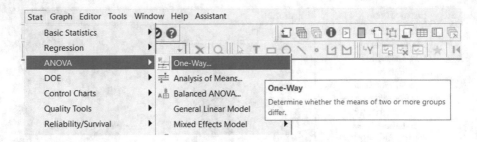

Fig. 5.32 Minitab path for ANOVA

5.5.2   ANOVA in Minitab

The Minitab path for ANOVA is Stat/ANOVA/One-Way (implying one response or Y), or as shown in Fig. 5.32. Also below are step-by-step instructions.

Step	Instructions	Illustration
1	• Paste the data from the ANOVA dataset into the Minitab worksheet. • Note the four columns of tensile strength results for each lot or type of cast iron alloy material.	<table><tr><th>C1</th><th>C2</th><th>C3</th><th>C4</th></tr><tr><th>Lot 1 failed calipers</th><th>Lot 2</th><th>Lot 3</th><th>Lot 4</th></tr><tr><td>409.594</td><td>412.598</td><td>413.595</td><td>416.909</td></tr><tr><td>406.018</td><td>424.459</td><td>417.886</td><td>404.425</td></tr><tr><td>439.926</td><td>419.445</td><td>423.270</td><td>418.407</td></tr><tr><td>391.406</td><td>398.386</td><td>420.456</td><td>395.683</td></tr><tr><td>403.661</td><td>408.199</td><td>409.561</td><td>419.951</td></tr></table>
2	Checking the normality assumption • Minitab path: Graph/probability plot. • Normality: Confirmed—All P-values > alpha 0.05.	**Probability Plot of Lot 1 failed calipers, Lot 2, Lot 3, Lot 4** Normal - 95% CI Lot 1 failed calipers — Mean 413.8, StDev 13.81, N 30, AD 0.296, P-Value 0.570 Lot 2 — Mean 417.8, StDev 10.56, N 30, AD 0.267, P-Value 0.665 Lot 3 — Mean 414.8, StDev 9.201, N 30, AD 0.616, P-Value 0.099 Lot 4 — Mean 413.1, StDev 8.625, N 30, AD 0.324, P-Value 0.510 Project: ; Worksheet: Worksheet 5; 11/6/2021

Step	Instructions	Illustration
3	Checking the equal variances assumption • Minitab path: Stat/ ANOVA/test for equal variances. • Select "response data are in a separate column for each factor level". • In "responses," select all four columns of data. • Select "options".	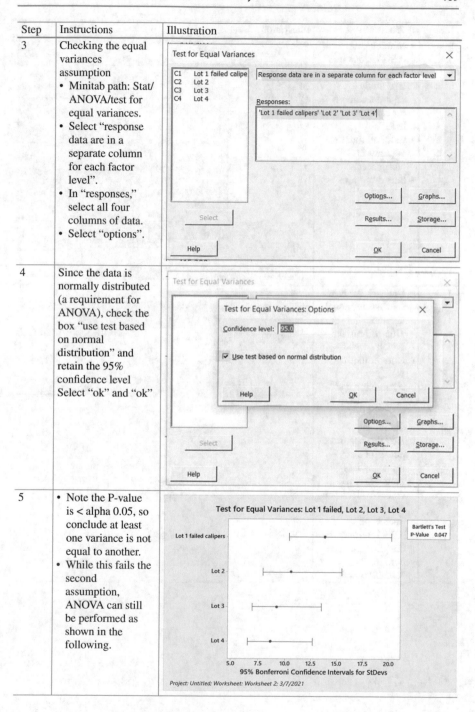
4	Since the data is normally distributed (a requirement for ANOVA), check the box "use test based on normal distribution" and retain the 95% confidence level Select "ok" and "ok"	
5	• Note the P-value is < alpha 0.05, so conclude at least one variance is not equal to another. • While this fails the second assumption, ANOVA can still be performed as shown in the following.	

Step	Instructions	Illustration
6	• Minitab path: Stat/ ANOVA/one way. • Select "response data are in a separate column for each factor level". • Enter all four data columns in "responses". • Select "options".	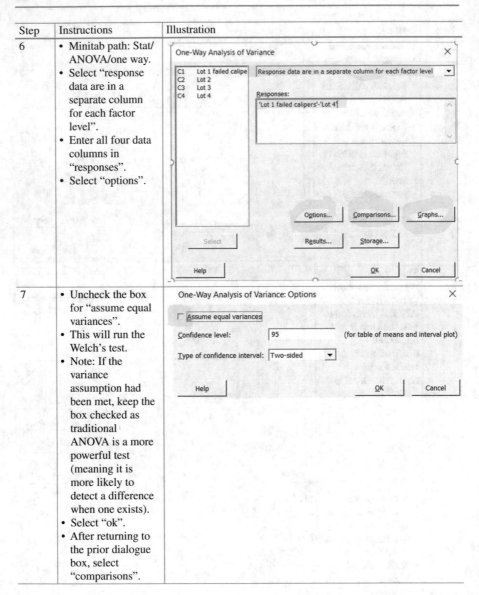
7	• Uncheck the box for "assume equal variances". • This will run the Welch's test. • Note: If the variance assumption had been met, keep the box checked as traditional ANOVA is a more powerful test (meaning it is more likely to detect a difference when one exists). • Select "ok". • After returning to the prior dialogue box, select "comparisons".	

Step	Instructions	Illustration
8	• Select the box beside "games-Howell" (note: If the "assume equal variance" box had been checked, select "Fishers"). • Select "ok". • After returning to the initial dialogue box, select "graphs".	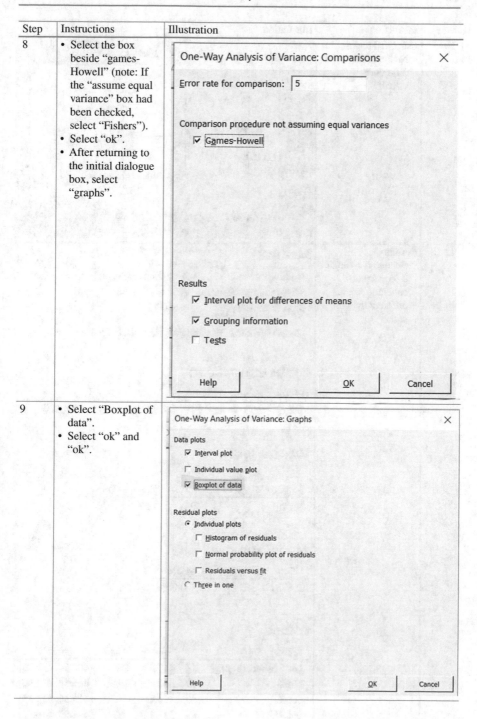
9	• Select "Boxplot of data". • Select "ok" and "ok".	

Step	Instructions	Illustration
10	Graphical results • No apparent difference visually.	 Project: Untitled; Worksheet: Worksheet 1; 2/20/2020
11	Results • P-value is > alpha 0.05; cannot conclude one is different than the other.	*(see analysis below)*

Method

Null hypothesis	All means are equal
Alternative hypothesis	Not all means are equal
Significance level	$\alpha = 0.05$

Equal variances were not assumed for the analysis.

Factor Information

Factor	Levels	Values
Factor	4	Lot 1 failed calipers, Lot 2, Lot 3, Lot 4

Welch's Test

Source	DF Num	DF Den	F-Value	P-Value
Factor	3	63.7400	1.22	0.309

Model Summary

R-sq	R-sq(adj)	R-sq(pred)
2.80%	0.29%	0.00%

Means

Factor	N	Mean	StDev	95% CI
Lot 1 failed calipers	30	413.84	13.81	(408.68, 418.99)
Lot 2	30	417.82	10.56	(413.88, 421.76)
Lot 3	30	414.81	9.20	(411.38, 418.25)
Lot 4	30	413.12	8.63	(409.90, 416.34)

Step	Instructions	Illustration
12	Determining which means are different • Fishers (when variances are considered equal) and games-Howell (when variances are unequal) show which means are different. • Means are considered different between any two samples if their confidence interval does not cross the zero line. • In this case all confidence intervals cross 0, so none can be considered different from any other.	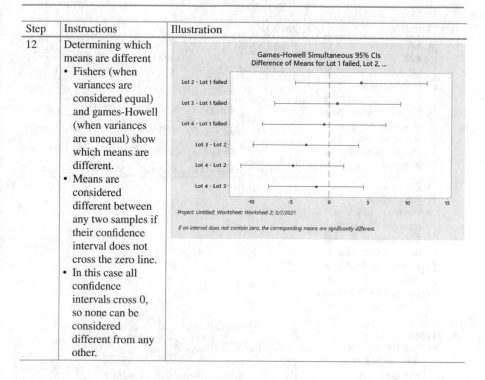

5.5.3 ANOM in Minitab

Another tool that is especially helpful visualizing differences between more than two sample means is analysis of means, or ANOM. The Minitab path for ANOM is Stat/ANOVA/Analysis of Means and as shown in Fig. 5.33. As noted prior, this test determines whether a mean is different than the mean of the means. (This is different from ANOVA, which determines if at least one mean is different than another.) Below are step-by-step instructions.

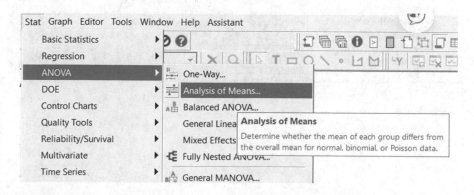

Fig. 5.33 Minitab path for ANOM

Step	Instructions	Illustration
1	• Paste the data from the ANOVA dataset into the Minitab worksheet (this is the same as for the ANOVA example). • Note the four columns of tensile strength results for each lot or type of alloy cast iron material.	<table><thead><tr><th>C1</th><th>C2</th><th>C3</th><th>C4</th></tr><tr><th>Lot 1 failed calipers</th><th>Lot 2</th><th>Lot 3</th><th>Lot 4</th></tr></thead><tbody><tr><td>409.594</td><td>412.598</td><td>413.595</td><td>416.909</td></tr><tr><td>406.018</td><td>424.459</td><td>417.886</td><td>404.425</td></tr><tr><td>439.926</td><td>419.445</td><td>423.270</td><td>418.407</td></tr><tr><td>391.406</td><td>398.386</td><td>420.456</td><td>395.683</td></tr><tr><td>403.661</td><td>408.199</td><td>409.561</td><td>419.051</td></tr></tbody></table>
2	• Check the normality assumption. • See the ANOVA example for how to do this, which confirmed the assumption for a normal distribution is met. Note: ANOM can also be used for binomial and Poisson data	
3	Stacking the columns • ANOM requires stacked data, or data in a single column. • Add two column headings, "stacked yield" and "lot". • Minitab path: Data/ stack/columns. • Select all four columns in "stack the following columns". • Select "column of current worksheet" and enter "stacked yield". • Enter "lot" in "store subscripts in". • Ensure the box is checked for "use variable names in subscript column". • Select "ok".	

Step	Instructions	Illustration
4	Performing the analysis • Minitab path: Stat/ ANOVA/analysis of means. • Enter "stacked yield" for "response". • Select "Normal". • Enter "lot" for "factor 1". • Select "OK".	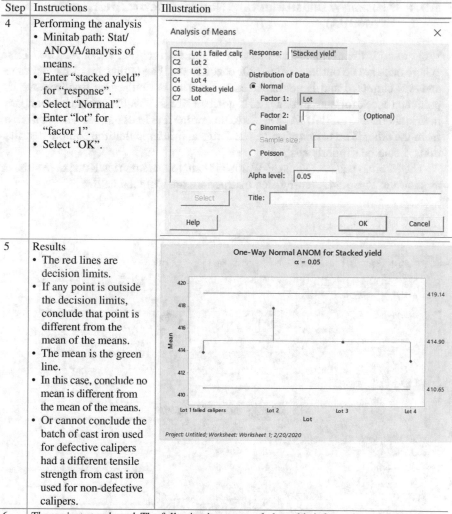
5	Results • The red lines are decision limits. • If any point is outside the decision limits, conclude that point is different from the mean of the means. • The mean is the green line. • In this case, conclude no mean is different from the mean of the means. • Or cannot conclude the batch of cast iron used for defective calipers had a different tensile strength from cast iron used for non-defective calipers.	
6	*The project storyboard*: The following is an example how this information might appear in a project storyboard:	

Storyboard: ANOM results for defective caliper cast iron tensile strength

- No points outside the Decision Limits
- Conclude: No mean is different from the mean of the means
- Conclude: Cannot conclude the batch of cast iron used for defective calipers had a different tensile strength from cast iron used for non-defective calipers

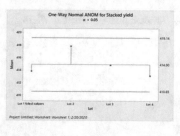

5.5.4   Kruskal-Wallis in Minitab (Nonparametric Alternative to ANOVA)

What to do if the normality assumption is not met for ANOVA or ANOM? First, ensure there are no outliers in error. Or check to see if an outlier was a special cause and was corrected and is not expected to reoccur. For either are true, remove the point and check for normality again. If not, run Kruskal-Wallis, which is a nonparametric alternative to ANOVA. It seeks to determine if at least one median is different from the other. The assumption to use Kruskal-Wallis is that the sample distributions should be similarly shaped.

The Minitab path for Kruskal-Wallis is Stat/Nonparametrics/Kruskal-Wallis, or as shown in Fig. 5.34. Step-by-step instructions are included below.

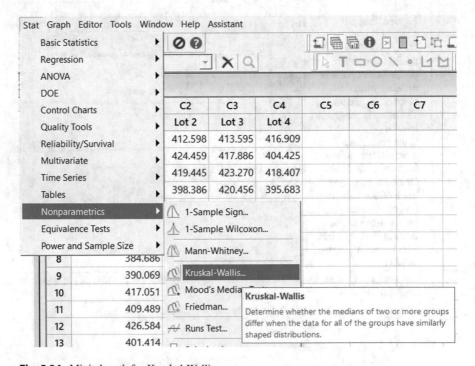

Fig. 5.34   Minitab path for Kruskal-Wallis

Step	Instructions	Illustration
1	• Paste the data from the ANOVA dataset into the Minitab worksheet. • Note the four columns of tensile strength results for each lot or type of alloy cast iron material. • Note this is the same data used in the previous ANOVA example.	
2	Stacking the columns • Kruskal Wallis requires stacked data, or data in a single column. • Add two column headings, "stacked yield" and "lot" (this was also done in the prior steps for ANOM). • Minitab path: Data/ stack/columns. • Select all four columns in "stack the following columns". • Select "column of current worksheet" and enter "stacked yield". • Enter "lot" in "store subscripts in". • Ensure the box is checked for "use variable names in subscript column". • Select "ok".	

Illustration for Step 1:

C1	C2	C3	C4
Lot 1 failed calipers	Lot 2	Lot 3	Lot 4
409.594	412.598	413.595	416.909
406.018	424.459	417.886	404.425
439.926	419.445	423.270	418.407
391.406	398.386	420.456	395.683
403.661	408.199	409.561	419.951

Illustration for Step 2:

↓	C1	C2	C3	C4	C5	C6
	Lot 1 failed calipers	Lot 2	Lot 3	Lot 4	Stacked Yield	Lot
1	409.594	412.598	413.595	416.909		
2	406.018	424.459	417.886	404.425		

Stack Columns ✕

Stack the following columns:

'Lot 1 failed calipers' 'Lot 2' 'Lot 3' 'Lot 4'

Store stacked data in:
○ New worksheet
   Name: (Optional)

● Column of current worksheet: 'Stacked Yield'
   Store subscripts in: Lot   (Optional)

☑ Use variable names in subscript column

Select

Help   OK   Cancel

Step	Instructions	Illustration
3	Performing the analysis • Minitab path: Stat/ Nonparametrics/ Kruskal-Wallis. • Enter "stacked yield" for "response". • Enter "lot" for "factor". • Select "ok".	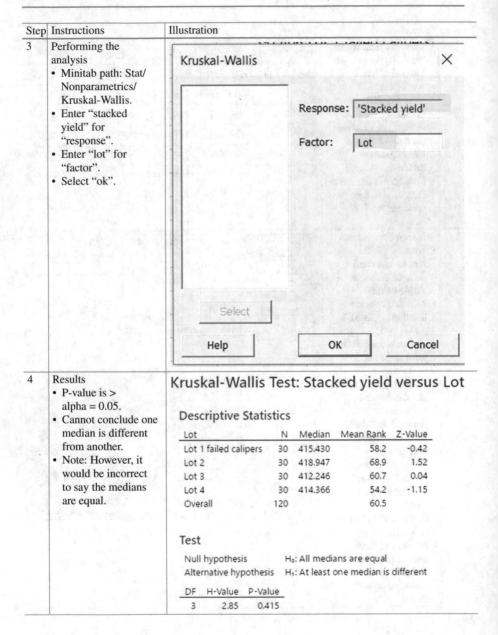
4	Results • P-value is > alpha = 0.05. • Cannot conclude one median is different from another. • Note: However, it would be incorrect to say the medians are equal.	**Kruskal-Wallis Test: Stacked yield versus Lot** **Descriptive Statistics** Lot ... N ... Median ... Mean Rank ... Z-Value Lot 1 failed calipers ... 30 ... 415.430 ... 58.2 ... -0.42 Lot 2 ... 30 ... 418.947 ... 68.9 ... 1.52 Lot 3 ... 30 ... 412.246 ... 60.7 ... 0.04 Lot 4 ... 30 ... 414.366 ... 54.2 ... -1.15 Overall ... 120 ... ... 60.5 **Test** Null hypothesis ... H_0: All medians are equal Alternative hypothesis ... H_1: At least one median is different DF ... H-Value ... P-Value 3 ... 2.85 ... 0.415

The Kruskal-Wallis illustration dialog reads:

Kruskal-Wallis ✕

Response: 'Stacked yield'

Factor: Lot

Select

Help   OK   Cancel

Step	Instructions	Illustration
5	Checking the similarly shaped distributions assumption • Minitab path: Graphs/histogram simple. • Select the original columns of the for lots of data. • Select "multiple graphs".	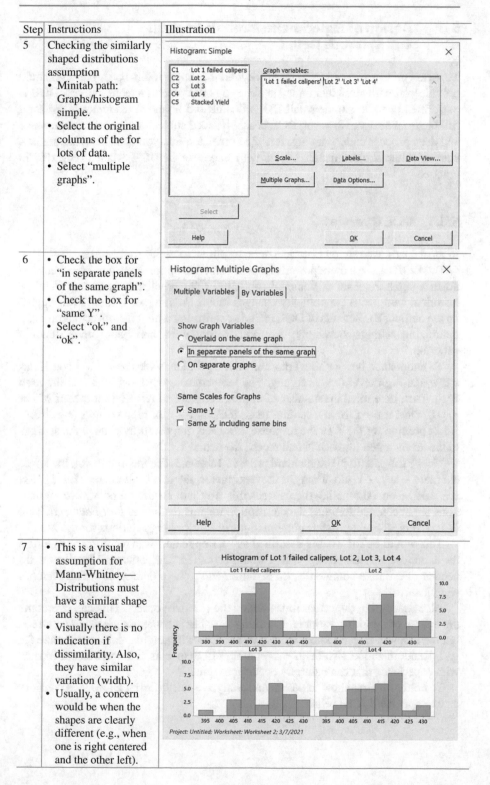
6	• Check the box for "in separate panels of the same graph". • Check the box for "same Y". • Select "ok" and "ok".	
7	• This is a visual assumption for Mann-Whitney— Distributions must have a similar shape and spread. • Visually there is no indication if dissimilarity. Also, they have similar variation (width). • Usually, a concern would be when the shapes are clearly different (e.g., when one is right centered and the other left).	

5.6   Design of Experiment: Minitab Analysis and Methods Detail

There are other advanced DOE topics beyond the scope of this book, but this should provide you with sufficient understanding to participate in a team where a DOE is performed or to do simple small DOEs if time and financial resources allow for a full factorial design. However, DOE is an advanced subject and requires experience and deep process understanding. It is recommended that you consult with a statistician or other subject matter expert with expertise in DOE when doing this in practice.

5.6.1   DOE Overview

In most cases, we rely on historical data, or pilots, to verify root causes.

But DOE can be a more powerful and informative option once you have funneled down a smaller subset of potential root causes (or predictors or factors). Unlike historical data, DOE intentionally changes inputs (X's) to see if there is an impact on the output (Y). Benefits of DOE include confirming which factors are significant, optimizing settings, developing a predictive equation, and better understanding variation.

To understand how DOE works, consider a problem with *three* X's. Each X has a low and high level (e.g., one might be temperature, with low 59F, and the high 77F). Then determine the number of possible combinations = 2^(number of X's or 3) = 8. Then use −1 for low settings (e.g., 59F) and + 1 for high settings (e.g., 77F). Add a column for the Y, or the response. For each run, carefully change the settings to the levels noted for each X and record the actual Y.

Notice the pattern in the standard order in Table 5.3. For the first factor, the levels alternate −1, 1, −1, etc. Then for the next factor, the levels alternate −1, −1, 1, 1, etc. And so on. Alternating the patterns this way will ensure all possible combinations are accounted for, thus the description for this method is *Full Factorial*. That is, it covers all eight possible combinations of low and high settings for the X's.

This would be for what is described as "1 Replicate." It is recommended to run the experiment at least twice, which would be called "2 Replicates." Replicates are needed to estimate common cause variation and to help determine which factor is significant.

Also randomize the actual runs (versus the standard order). This can overcome the effects of a lurking variable. A lurking variable is something causing an effect, but the assessor doesn't know what it is. Minitab will automatically randomize for you. Randomization also helps overcoming effect of bias or improvement in runs with experience that can naturally occur over time.

To better illustrate how to design and analyze a full factorial DOE, let's consider an example from KIND Karz.

Table 5.3   Example full factorial DOE design with 1 replicate

Worksheet 1 ***				
♦	C1	C2	C3	C4	C5
	StdOrder	X1	X2	X3	Y
1	1	-1	-1	-1	
2	2	1	-1	-1	
3	3	-1	1	-1	
4	4	1	1	-1	
5	5	-1	-1	1	
6	6	1	-1	1	
7	7	-1	1	1	
8	8	1	1	1	

5.6.2   Full Factorial DOE Example in Minitab (No Interactions Between X's)

Prior in this book, we confirmed the new assembler was leading to more failures in airbags. But why is there a difference? After further process analysis, Albert reviewed the new machine with the vendor technician. After comparing the new machine to others, it was discovered there were new settings that needed to be set differently from the other assembler to ensure an adequate seal strength of the airbag. What are the optimum settings to ensure the seal strength is sufficient? DOE can help determine this, as will be shown in the next step-by-step instructions.

For this problem, the response will be seal strength measured in PSI. The team has narrowed down a list of most likely predictors as follows:

- Temperature (low 180, high 200).
- Vacuum (low 30, high 50).
- Time (low 100 milliseconds, high 200 milliseconds).
- Glue type (low Brand A, high Brand B)—also will help us determine if glue is a determining factor.

As noted prior, we need to have at least 2 replicates (meaning run the DOE twice). So our number of runs = 2 replicates * 2 levels ^(4 factors) = 32 runs. But Minitab can help us properly set up the design (Stat/DOE/Factorial/Create Factorial Design) as shown in Table 5.3. Later, we will see how Minitab can make the analysis easy (Stat/DOE/Factorial/Analyze Factorial Design) (Figs. 5.35 and 5.36).

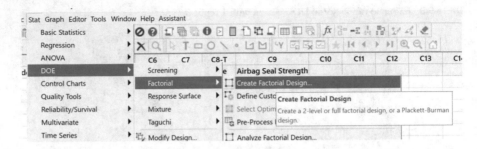

Fig. 5.35 Minitab path to design a DOE

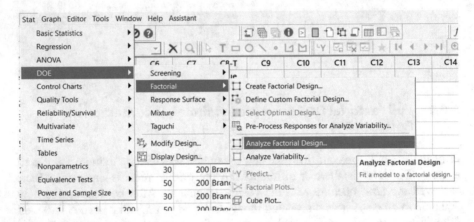

Fig. 5.36 Minitab path to analyzing a DOE

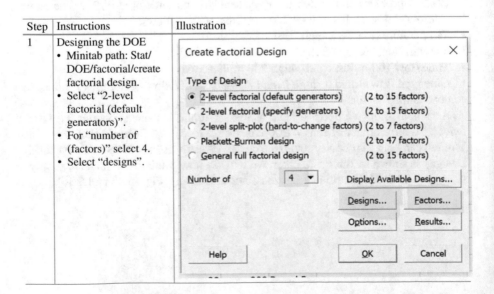

Step	Instructions	Illustration
1	Designing the DOE • Minitab path: Stat/ DOE/factorial/create factorial design. • Select "2-level factorial (default generators)". • For "number of (factors)" select 4. • Select "designs".	

Step	Instructions	Illustration
2	• Select "full factorial". • For "number of center points per block," select 0. • For "number of replicates for corner," select 2. • For "number of blocks," select 1. • Select ok. • After returning to the prior dialogue box, select "factors". Note on blocking: Blocking is used to explain variation not directly caused by the factors. For example, a DOE might span equipment, shifts, or plant locations. If there is concern, these could be contributing factors, and consider adding blocks. However, be aware the study resolution may decrease. Resolution will be discussed later	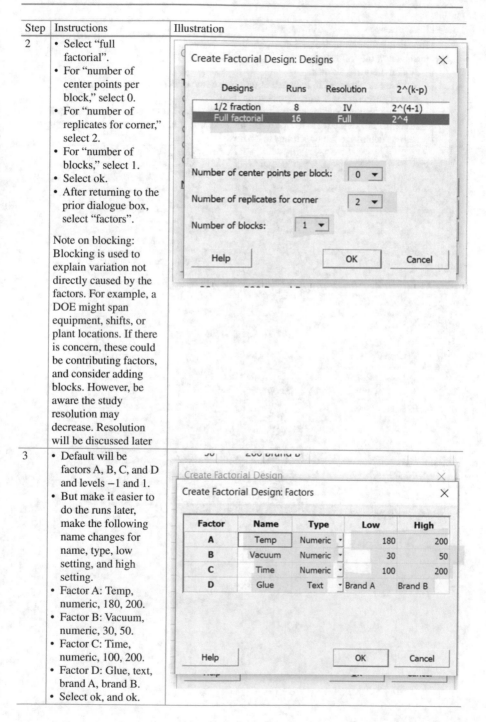
3	• Default will be factors A, B, C, and D and levels −1 and 1. • But make it easier to do the runs later, make the following name changes for name, type, low setting, and high setting. • Factor A: Temp, numeric, 180, 200. • Factor B: Vacuum, numeric, 30, 50. • Factor C: Time, numeric, 100, 200. • Factor D: Glue, text, brand A, brand B. • Select ok, and ok.	

Step	Instructions	Illustration
4	Reviewing the completed design • Shown in the following. • Notice there are 32 runs. • Runs are randomized to reduce bias and overcome effects of lurking variables. • Add column for the Y (airbag seal strength). The next step is to change the settings 32 times (i.e. 32 runs) and measure the seal strength. The objective is to maximize the seal strength, and the analysis step later will help us determine factor settings to do this	

▪	C1 StdOrder	C2 RunOrder	C3 CenterPt	C4 Blocks	C5 Temp	C6 Vacuum	C7 Time	C8-T Glue	C9 Airbag Seal Strength
1	2	1	1	1	200	30	100	Brand A	
2	19	2	1	1	180	50	100	Brand A	
3	15	3	1	1	180	50	200	Brand B	
4	20	4	1	1	200	50	100	Brand A	
5	32	5	1	1	200	50	200	Brand B	
6	5	6	1	1	180	30	200	Brand A	
7	4	7	1	1	200	50	100	Brand A	
8	21	8	1	1	180	30	200	Brand A	
9	17	9	1	1	180	30	100	Brand A	
10	31	10	1	1	180	50	200	Brand B	
11	24	11	1	1	200	50	200	Brand A	
12	1	12	1	1	180	30	100	Brand A	
13	22	13	1	1	200	30	200	Brand A	
14	6	14	1	1	200	30	200	Brand A	
15	29	15	1	1	180	30	200	Brand B	
16	14	16	1	1	200	30	200	Brand B	
17	12	17	1	1	200	50	100	Brand B	
18	10	18	1	1	200	30	100	Brand B	
19	18	19	1	1	200	30	100	Brand A	
20	30	20	1	1	200	30	200	Brand B	
21	11	21	1	1	180	50	100	Brand B	
22	28	22	1	1	200	50	100	Brand B	
23	25	23	1	1	180	30	100	Brand B	
24	27	24	1	1	180	50	100	Brand B	
25	16	25	1	1	200	50	200	Brand B	
26	13	26	1	1	180	30	200	Brand B	
27	9	27	1	1	180	30	100	Brand B	
28	23	28	1	1	180	50	200	Brand A	
29	3	29	1	1	180	50	100	Brand A	
30	8	30	1	1	200	50	200	Brand A	
31	7	31	1	1	180	50	200	Brand A	
32	26	32	1	1	200	30	100	Brand B	

Step	Instructions	Illustration
5	• Then perform the tests in the random order and enter the Y's in the RunOrder. • If you would like to follow for this example, refer to the DOE dataset. (for purposes of following along, delete the rows in Minitab but keep the headers and paste the rows from the dataset). • Also note your randomized order might be different. • Also note there will be an additional response column for later analysis that will demonstrate interactions between X's.	(see table below)

StdOrder	RunOrder	CenterPt	Blocks	Temp	Vacuum	Time	Glue	Airbag Seal Strength	Airbag Seal Strength with Interaction
2	1	1	1	200	30	100	Brand A	1090	1390
19	2	1	1	180	50	100	Brand A	929	1379
15	3	1	1	180	50	200	Brand B	1010	1460
20	4	1	1	200	50	100	Brand A	1027	1527

Step	Instructions	Illustration
6	Perform the analysis • Minitab path: Stat/DOE/factorial/Analyze factorial design. • In "responses," select "airbag seal strength". • Select "graphs".	(Analyze Factorial Design dialog box)

Analyze Factorial Design   ✕

C9   Airbag Seal Strength   Responses:
C10 Airbag Seal Strengtl   'Airbag Seal Strength'

Terms...   Covariates...   Options...   Stepwise...
Graphs...   Results...   Storage...

Select

Help   OK   Cancel

Step	Instructions	Illustration
7	• Select "Pareto" and "four in one". • Select ok and ok.	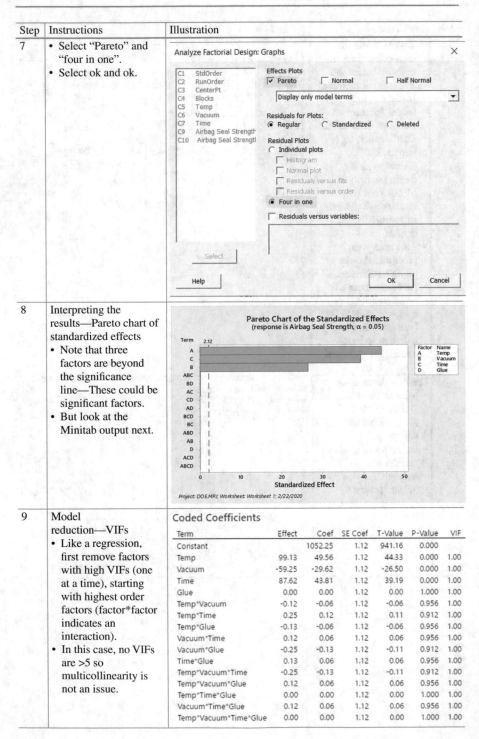
8	Interpreting the results—Pareto chart of standardized effects • Note that three factors are beyond the significance line—These could be significant factors. • But look at the Minitab output next.	
9	Model reduction—VIFs • Like a regression, first remove factors with high VIFs (one at a time), starting with highest order factors (factor*factor indicates an interaction). • In this case, no VIFs are >5 so multicollinearity is not an issue.	

Coded Coefficients

Term	Effect	Coef	SE Coef	T-Value	P-Value	VIF
Constant		1052.25	1.12	941.16	0.000	
Temp	99.13	49.56	1.12	44.33	0.000	1.00
Vacuum	-59.25	-29.62	1.12	-26.50	0.000	1.00
Time	87.62	43.81	1.12	39.19	0.000	1.00
Glue	0.00	0.00	1.12	0.00	1.000	1.00
Temp*Vacuum	-0.12	-0.06	1.12	-0.06	0.956	1.00
Temp*Time	0.25	0.12	1.12	0.11	0.912	1.00
Temp*Glue	-0.13	-0.06	1.12	-0.06	0.956	1.00
Vacuum*Time	0.12	0.06	1.12	0.06	0.956	1.00
Vacuum*Glue	-0.25	-0.13	1.12	-0.11	0.912	1.00
Time*Glue	0.13	0.06	1.12	0.06	0.956	1.00
Temp*Vacuum*Time	-0.25	-0.13	1.12	-0.11	0.912	1.00
Temp*Vacuum*Glue	0.12	0.06	1.12	0.06	0.956	1.00
Temp*Time*Glue	0.00	0.00	1.12	0.00	1.000	1.00
Vacuum*Time*Glue	0.12	0.06	1.12	0.06	0.956	1.00
Temp*Vacuum*Time*Glue	0.00	0.00	1.12	0.00	1.000	1.00

Step	Instructions	Illustration
10	Reducing the model—P-values • Remove factors or interaction terms with P-values >0.05 one at a time, starting with highest order interactions. • Start with four-way interactions for this example. • Note the P-value is high (greater than alpha 0.05) for four-way interactions. • Remove from the model and rerun.	**Analysis of Variance**

Analysis of Variance

Source	DF	Adj SS	Adj MS	F-Value	P-Value
Model	15	168118	11207.9	280.20	0.000
Linear	4	168116	42028.9	1050.72	0.000
Temp	1	78606	78606.1	1965.15	0.000
Vacuum	1	28085	28084.5	702.11	0.000
Time	1	61425	61425.1	1535.63	0.000
Glue	1	0	0.0	0.00	1.000
2-Way Interactions	6	2	0.3	0.01	1.000
Temp*Vacuum	1	0	0.1	0.00	0.956
Temp*Time	1	0	0.5	0.01	0.912
Temp*Glue	1	0	0.1	0.00	0.956
Vacuum*Time	1	0	0.1	0.00	0.956
Vacuum*Glue	1	1	0.5	0.01	0.912
Time*Glue	1	0	0.1	0.00	0.956
3-Way Interactions	4	1	0.2	0.00	1.000
Temp*Vacuum*Time	1	1	0.5	0.01	0.912
Temp*Vacuum*Glue	1	0	0.1	0.00	0.956
Temp*Time*Glue	1	0	0.0	0.00	1.000
Vacuum*Time*Glue	1	0	0.1	0.00	0.956
4-Way Interactions	1	0	0.0	0.00	1.000
Temp*Vacuum*Time*Glue	1	0	0.0	0.00	1.000
Error	16	640	40.0		
Total	31	168758			

Step	Instructions	Illustration
11	• Rerun the analysis or ctrl-E to relaunch the DOE dialogue box. • Select "terms".	

Analyze Factorial Design

C9 Airbag Seal Strength

Responses:

Airbag Seal Strength

Terms...   Covariates...   Options...   Stepwise...

Graphs...   Results...   Storage...

Select

Help   OK   Cancel

Step	Instructions	Illustration
12	• Change the "include terms in the model up through" to 3. • This will eliminate the four-way interaction. • Keep doing this model reduction step until only interactions with P-values less than 0.05 remains. • You will notice the three-way interactions have P-values greater than 0.05. • Reduce to 2, and you will notice the same. • Then reduce to 1. • Note: If any of the interactions within a common group of interactions has a P-value less than 0.05, then the number of terms in the model cannot be reduced, and factors will need to be removed one at a time.	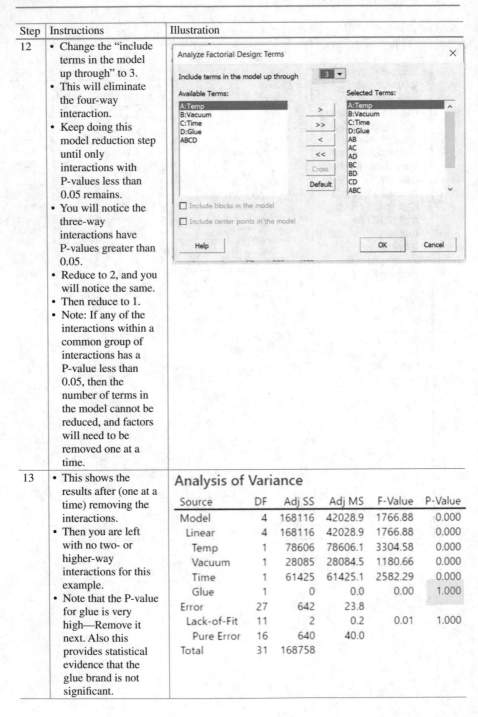

| 13 | • This shows the results after (one at a time) removing the interactions.
• Then you are left with no two- or higher-way interactions for this example.
• Note that the P-value for glue is very high—Remove it next. Also this provides statistical evidence that the glue brand is not significant. | **Analysis of Variance**

See table below |

Analysis of Variance

Source	DF	Adj SS	Adj MS	F-Value	P-Value
Model	4	168116	42028.9	1766.88	0.000
Linear	4	168116	42028.9	1766.88	0.000
Temp	1	78606	78606.1	3304.58	0.000
Vacuum	1	28085	28084.5	1180.66	0.000
Time	1	61425	61425.1	2582.29	0.000
Glue	1	0	0.0	0.00	1.000
Error	27	642	23.8		
Lack-of-Fit	11	2	0.2	0.01	1.000
Pure Error	16	640	40.0		
Total	31	168758			

Step	Instructions	Illustration
14	• Rerun or ctrl-E. • Notice now the reduced model has a 1 for "include terms in the model up through order" as you defined in your last run. • Click on the factor you would like to remove and then the < next to it. • Click "ok" and "ok".	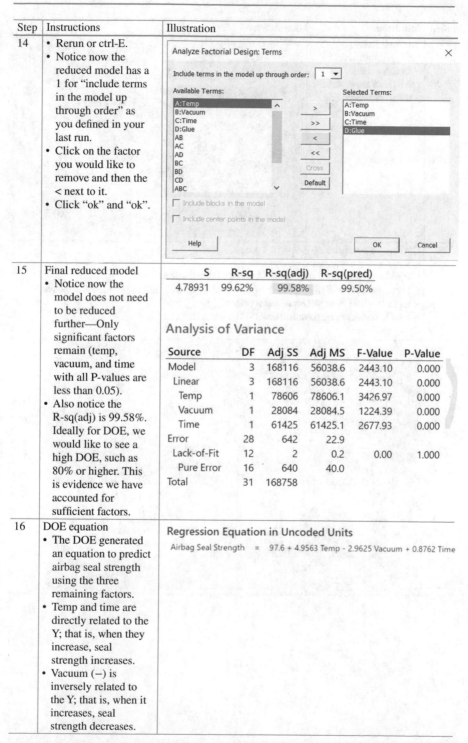

15	Final reduced model • Notice now the model does not need to be reduced further—Only significant factors remain (temp, vacuum, and time with all P-values are less than 0.05). • Also notice the R-sq(adj) is 99.58%. Ideally for DOE, we would like to see a high DOE, such as 80% or higher. This is evidence we have accounted for sufficient factors.	S R-sq R-sq(adj) R-sq(pred) 4.78931 99.62% 99.58% 99.50% ## Analysis of Variance | Source	DF	Adj SS	Adj MS	F-Value	P-Value
Model	3	168116	56038.6	2443.10	0.000
Linear	3	168116	56038.6	2443.10	0.000
Temp	1	78606	78606.1	3426.97	0.000
Vacuum	1	28084	28084.5	1224.39	0.000
Time	1	61425	61425.1	2677.93	0.000
Error	28	642	22.9		
Lack-of-Fit	12	2	0.2	0.00	1.000
Pure Error	16	640	40.0		
Total	31	168758				

| 16 | DOE equation
• The DOE generated an equation to predict airbag seal strength using the three remaining factors.
• Temp and time are directly related to the Y; that is, when they increase, seal strength increases.
• Vacuum (−) is inversely related to the Y; that is, when it increases, seal strength decreases. | ## Regression Equation in Uncoded Units

Airbag Seal Strength = 97.6 + 4.9563 Temp - 2.9625 Vacuum + 0.8762 Time |

Step	Instructions	Illustration
17	The final Pareto • Only significant factors remain. • All three are beyond the significance line.	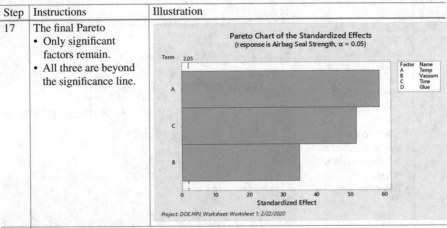
18	Checking assumptions—See 4-in-1 graph below • Same as for a regression analysis. • Residuals meet the requirement for normal distribution. • No pattern of residuals versus fits (the apparent mirror pattern is because we have replicates and does not represent a pattern). • Residuals versus observation order—No concern.	

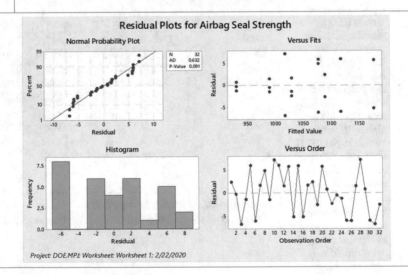

Step	Instructions	Illustration
19	Optimal setting—Cube plot • Generate a cube plot (which works in this case since we have three predictors). • Minitab path: Stat/DOE/factorial/cube plot. • But intuitively, from seeing the equation, we would expect high temperature and time with low vacuum. • The cube plot confirms this.	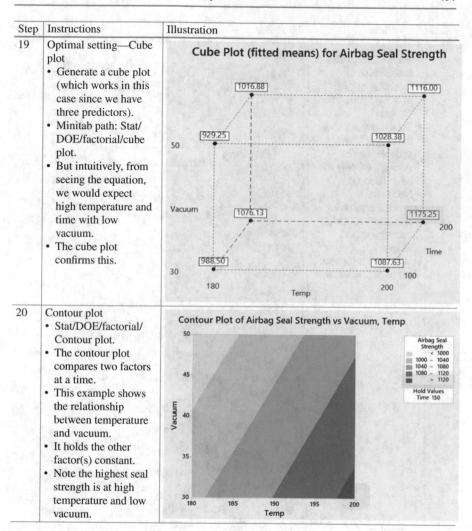
20	Contour plot • Stat/DOE/factorial/Contour plot. • The contour plot compares two factors at a time. • This example shows the relationship between temperature and vacuum. • It holds the other factor(s) constant. • Note the highest seal strength is at high temperature and low vacuum.	

Step	Instructions	Illustration
21	Surface plot • Stat/DOE/factorial/ surface plot. • Select "temp" for "X axis" and "time" for "Y Axis". • This example shows the effect of temp and time on seal strength, holding the other factor(s). • Notice the highest seal strength is at a combination of high temperature and longer time, consistent with what we have seen so far.	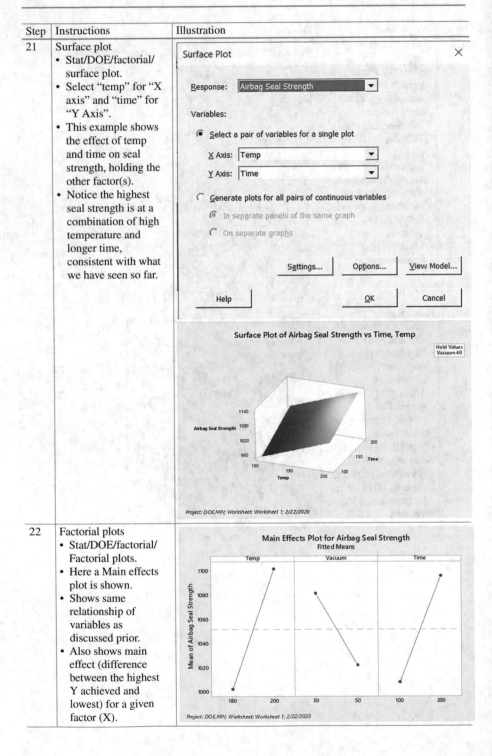 Project: DOE.MPJ; Worksheet: Worksheet 1; 2/22/2020
22	Factorial plots • Stat/DOE/factorial/ Factorial plots. • Here a Main effects plot is shown. • Shows same relationship of variables as discussed prior. • Also shows main effect (difference between the highest Y achieved and lowest) for a given factor (X).	

5.6.3 Full Factorial DOE Example in Minitab (with Interactions)

The prior example was for when there was no interaction between terms. The following uses a different possible outcome to illustrate what might look different and how to interpret when there is an interaction. These steps assume the prior analysis was run.

Step	Instructions	Illustration
1	• Notice in the dataset there is another column for Y, "airbag seal strength with interaction". • Rerun the analysis to illustrate what might occur if there were to be statistically significant interactions in the model. • Individual steps will not be repeated here for analyzing and reducing the model. • Note: You should review the prior example with no interaction before following along with this example.	<table><tr><th>StdOrder</th><th>RunOrder</th><th>CenterPt</th><th>Blocks</th><th>Temp</th><th>Vacuum</th><th>Time</th><th>Glue</th><th>Airbag Seal Strength</th><th>Airbag Seal Strength with Interaction</th></tr><tr><td>2</td><td>1</td><td>1</td><td>1</td><td>200</td><td>30</td><td>100</td><td>Brand A</td><td>1090</td><td>1390</td></tr><tr><td>19</td><td>2</td><td>1</td><td>1</td><td>180</td><td>50</td><td>100</td><td>Brand A</td><td>929</td><td>1379</td></tr><tr><td>15</td><td>3</td><td>1</td><td>1</td><td>180</td><td>50</td><td>200</td><td>Brand B</td><td>1010</td><td>1460</td></tr><tr><td>20</td><td>4</td><td>1</td><td>1</td><td>200</td><td>50</td><td>100</td><td>Brand A</td><td>1027</td><td>1527</td></tr></table>
2	Run the analysis • Minitab path: Stat/ DOE/factorial/analyze factorial design. • Reduce the model one factor at a time as described in the prior example. • When you get to the model reduced to include two-way interactions, you will notice temp*vacuum will have a P-value <0, so the entire category cannot be removed. Start removing one factor at a time starting with the highest two-way interactions. • What is shown here is the final reduced model. • Note the three significant factors remain along with the interaction of temp and vacuum.	**Model Summary** | S	R-sq	R-sq(adj)	R-sq(pred) | |---|---|---|---| | 4.87672 | 99.82% | 99.79% | 99.75% | **Analysis of Variance** | Source	DF	Adj SS	Adj MS	F-Value	P-Value | |---|---|---|---|---|---| | Model | 4 | 353816 | 88454 | 3719.30 | 0.000 | | Linear | 3 | 353036 | 117679 | 4948.14 | 0.000 | | Temp | 1 | 154846 | 154846 | 6510.95 | 0.000 | | Vacuum | 1 | 136765 | 136765 | 5750.66 | 0.000 | | Time | 1 | 61425 | 61425 | 2582.80 | 0.000 | | 2-Way Interactions | 1 | 780 | 780 | 32.80 | 0.000 | | Temp*Vacuum | 1 | 780 | 780 | 32.80 | 0.000 | | Error | 27 | 642 | 24 | | | | Lack-of-Fit | 11 | 2 | 0 | 0.00 | 1.000 | | Pure Error | 16 | 640 | 40 | | | | Total | 31 | 354458 | | | |

Step	Instructions	Illustration
3	Reduced model Pareto • Note the interaction term. • Explanation of interaction: Must know one setting to determine the effect of another. E.g. short person might read a dial gauge with different results than a tall person. • Notice for the reduced model all three factors and AB (temp*glue) interaction remain significant.	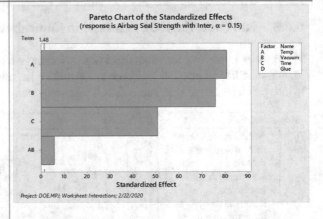

Pareto Chart of the Standardized Effects
(response is Airbag Seal Strength with Inter, α = 0.15)

Project: DOE.MPJ; Worksheet: Interactions; 2/22/2020

4	Checking assumptions—See 4-in-1 graph below

• Same as for a regression analysis.
• Meets the assumption of normally distributed residuals.
• No pattern of residuals versus fits (the apparent mirror pattern is because we have replicates and does not represent a pattern).
• Residuals versus observation order—No concern.

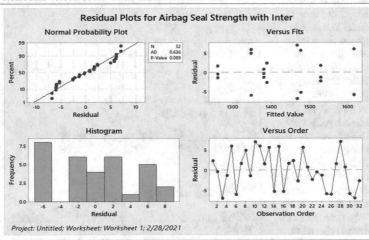

Residual Plots for Airbag Seal Strength with Inter

Project: Untitled; Worksheet: Worksheet 1; 2/28/2021

| 5 | Optimal setting—Cube plot
• Interestingly, the outcome changed.
• If there is an interaction, in this case, high time, temp, and vacuum are preferred. | 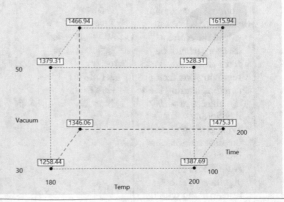 |

Cube Plot (fitted means) for Airbag Seal Strength with Inter

Step	Instructions	Illustration
6	Factorial plots • Note the main effects plot looks the same (not repeated here). • But now a new plot appears—The interaction plot. • To determine settings, one must consider both temp and vacuum together. • Sometimes these cross, affecting the selection. • In this case, in the set range of temperature, the lines don't cross so vacuum at 50 always gives a better result.	**Interaction Plot for Airbag Seal Strength with Inter** Fitted Means *Temp * Vacuum* Mean of Airbag Seal Strength with Inter Vacuum: 30.0, 50.0 Temp: 180.0, 200.0 *Project: DOE.MPJ; Worksheet: Interactions; 2/22/2020*
7	Factorial plots—What could happen • This is not an output of this dataset, but the graph is modified for illustration purposes. This is interpreted below: • If temp is 180 (low), then vacuum should be 500 (high) to achieve a higher seal strength. • If temp is 200 (high), then vacuum should be 300 (low).	**Interaction Plot for Airbag Seal Strength with Inter** Fitted Means *Temp * Vacuum* Mean of Airbag Seal Strength with Inter Vacuum: 30.0, 50.0 Temp: 180.0, 200.0 *Project: DOE.MPJ; Worksheet: Interactions; 2/22/2020*
8	*The project storyboard*: The following is an example how this information might appear in a project storyboard (with the interaction example):	

KIND Karz Storyboard: DOE, Assembler Sealing Strength

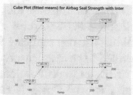

• Temp, Vacuum and Time are significant factors

• Standardize settings for high levels (Temp 200, Vacuum 50, and time 200)

Analyze

© Timothy Dean Blackburn 2020

5.6.4   Introduction to Screening Designs and Reduced Factorials

This section will introduce screening designs and reduced factorials. With many factors (Xs), the number of runs increase dramatically, increasing cost and time to implement. Minitab offers options to first screen which factors are likely significant but with less runs, including partial factorials, Plackett-Burman, and Definitive Screening Designs.

These methods dramatically reduce the number of runs needed but do result in a lower resolution study. That is, when all the combinations of low and high for the X's, one cannot tell what factor or interaction caused the effect. This is called confounding. When main effects or interactions are confounded, it is also called *aliased* with another effect. The main resolutions for DOE designs are indicated by Roman numerals and are as follows.

- Resolution III: Main effects (single X's) are aliased with two factor interactions. However, main effects are not confounded with other main effects.
- Resolution IV: Some two factor interactions are aliased with other two factor interactions. Also, main effects are aliased with three factor interactions. However, main effects are not aliased with another main effect, nor with a two-factor interaction.
- Resolution V: Two factor interactions are aliased with three factor interactions, and four factor interactions are aliased with main effects. However, main effects are not aliased with other main effects or with two factor interactions, nor are two factor interactions confounded with another two-factor interaction.

In practice, three factor interactions are rare, so usually Resolution IV DOEs are sufficient for screening and in some cases for final designs. However, for screening purposes (to narrow down significant factors before moving to a higher resolution DOE), Resolution III designs are often sufficient and have historically been used.

Minitab offers two specific reduced design tools—Plackett-Burman or Definitive Screening Design. The Minitab path is Stat/DOE/Screening/Create Screening Design. This is also shown in Fig. 5.35. A Placket-Burman can yield a Resolution III (no main effects are aliased with any other main effect, but main effects are aliased with two-factor interactions, which is risky unless used just for screening). However, the newer Definitive Screening Design method yields Resolution IV design (main effects are not aliased with any two-way interactions) (Fig. 5.37).

In addition to the screening designs, fractional factorial designs can also be created like what was described prior for full factorial. To illustrate this, imagine if we had four factors and ran a half fractional (full fractional would be 2^4 or 16 runs). To keep it simple and understandable, we will just use one replicate for purposes of illustration (although it is generally recommended to have at least two replicates).

In this example, it would be 8 runs (or ½ of 16, thus half fractional). To try this, the Minitab path is Stat/DOE/Factorial/Create Factorial Design/4 factors/Designs/½ Fraction. Click on "Display Available Designs" and see the resolution table shown in Fig. 5.38. Note the resolution in this case is IV for four factors and eight runs. Click "Ok" to see the design.

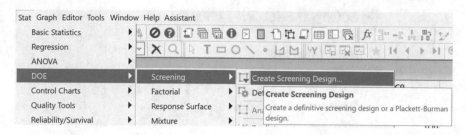

Fig. 5.37 Minitab screening design path

Create Factorial Design: Display Available Designs   ✕

Available Factorial Designs (with Resolution)

						Factors								
Runs	2	3	4	5	6	7	8	9	10	11	12	13	14	15
4	Full	III												
8		Full	IV	III	III	III								
16			Full	V	IV	IV	IV	III	III	III	III	III	III	III
32				Full	VI	IV	IV	IV	IV	IV	IV	IV	IV	IV
64					Full	VII	V	IV	IV	IV	IV	IV	IV	IV
128						Full	VIII	VI	V	V	IV	IV	IV	IV

Fig. 5.38 DOE fractional factorial resolutions

After creating the design, see the Minitab output information as shown in Fig. 5.39. It also indicates the resolution is IV. Further, it includes the Alias structure. Note, for example, two-way interactions are aliased with other two-way and single factors are aliased with three-way interactions. That is what we would expect for a resolution IV.

However, in some cases, choosing a half-fractional DOE can be a wise choice, because if only one of the factors (X) is determined not to be statistically significant, the analysis collapses to a full factorial. To illustrate this, let's continue with the half fractional study with four factors. The design is shown in Fig. 5.40. The factors are described as A, B, C, and D. However, imagine if factor D was eliminated during the analysis. Then the study would be left with the design as shown in Fig. 5.41. Note now the pattern is the same as for a full factorial, accounting for every possible combination of the three remaining factors.

To illustrate and compare the various methods, let's consider another example. Imaging we have narrowed down our factors to eight during the Measure phase using other (non-DOE) techniques. However, we are not certain if some of the factors are significant and having eight factors requires a lot of runs which leads to high cost, interruption to the business potentially, and added time. Again for sake of simplicity, let's assume one replicate (in practice we would have at least two) and compare the various options.

Fractional Factorial Design

Design Summary

Factors:	4	Base Design:	4, 8	Resolution:	IV
Runs:	8	Replicates:	1	Fraction:	1/2
Blocks:	1	Center pts (total):	0		

Design Generators: D = ABC

Alias Structure

I + ABCD
A + BCD
B + ACD
C + ABD
D + ABC
AB + CD
AC + BD
AD + BC

Fig. 5.39 Fractional factorial resolution and Alias structure

Fig. 5.40 Half fractional
factorial design

A	B	C	D
-1	-1	-1	-1
1	-1	-1	1
-1	1	-1	1
1	1	-1	-1
-1	-1	1	1
1	-1	1	-1
-1	1	1	-1
1	1	1	1

- Option 1: Full factorial: How many runs (just with one replicate) for a full factorial? $2^8 = \mathbf{256}$ runs.
- Option 2: Fractional Factorial: For a resolution IV factorial: **16** runs (Stat/DOE/ Factorial/Create a Factorial Design/Designs). In this case, we can run a 1/16 fractional factorial and still achieve a resolution IV (see Fig. 5.42).
- Option 3: Placket-Burman resolution III: **12** runs (Stat/DOE/Screening/Create Screening Design/Placket-Burman/Display Available Designs). Here, we can get the resolution III with only 12 runs—see Fig. 5.43.

Fig. 5.41 Collapse to a
full factorial

A	B	C
-1	-1	-1
1	-1	-1
-1	1	-1
1	1	-1
-1	-1	1
1	-1	1
-1	1	1
1	1	1

Fractional Factorial Design

Design Summary

Factors:	8	Base Design:	8, 16	Resolution:	IV
Runs:	16	Replicates:	1	Fraction:	1/16
Blocks:	1	Center pts (total):	0		

Fig. 5.42 Fractional factorial example

Fig. 5.43 Plackett-
Burman design example

Plackett-Burman Design

Design Summary

Factors:	8	Replicates:	1
Base runs:	12	Total runs:	12
Base blocks:	1	Total blocks:	1

- Option 4: Definitive Screening resolution IV: **17** runs (Stat/DOE/Screening/Create Screening Design/Definitive Screening/Display Available Designs). With just a few more runs than the Plackett-Burman approach, we can achieve a resolution IV (Fig. 5.44).

So which to choose? It really depends on how many factors, process knowledge, time, money available, etc. Consulting with a statistician is generally advised, even for Black Belts. Usually doing a full factorial at the start is not advised unless the process is very well known and there are few predictors (Xs). As shown above, we

Fig. 5.44 Definitive
screening design example

Definitive Screening Design

Design Summary

Factors:	8	Replicates:	1
Base runs:	17	Total runs:	17
Base blocks:	1	Total blocks:	1
Center points:	1		

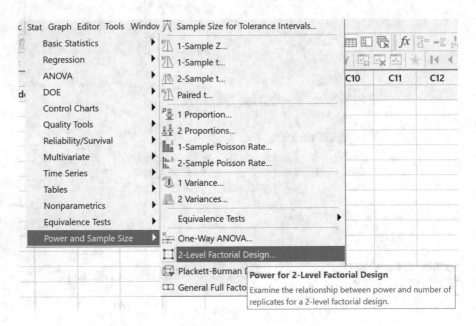

Fig. 5.45 Power and sample size for DOE

can still get a resolution IV with 16 or 17 runs. And remember the design may collapse to higher resolution as factors become insignificant, perhaps leading to a decision to use the fractional factorial in this case.

5.6.5 Other DOE Concepts and Methods

To aid in determining how many replicates you will need to have a significant DOE outcome if it exists, Minitab provides a tool to help with this. The Minitab path is Stat/Power and Sample Size/2-Level Factorial Design or see Fig. 5.45. Use a power value of 80% and estimate the effect level you wish the model capable of detecting. Start with assuming 0 center points per block. Corner points is the number of runs when factors are set a high and low levels and when at 1 replicate. You will need some historical data of the process to estimate standard deviation of the response.

There are other more advanced DOE techniques that are beyond the scope of the book, but it is helpful to be aware of these. For example, consider a situation where the model is related to two response variables (or Y's). In this case, Minitab includes an optimizer tool for this (Stat/DOE/Factorial/Response Optimizer). To use this, run the DOE analysis first with the two Y's.

Another advanced DOE technique is Response Surface (Stat/DOE/Response Surface/Create Response Surface Design). Use this when you expect curvature, typically after other DOEs have already identified significant factors. This helps to optimize the response better and is useful when there are squared (or quadratic) terms.

Again, DOE is an advanced topic in general, but this section has provided an overview at a sufficient level to perhaps allow you to lead full factorial designs when factors are few and the process is well understood, and some screening designs, although it is recommended to partner with someone more experienced the first few times at least. DOEs are very helpful but are also costly and time-consuming. It is important to *get it right* the first time.

References

Minitab Online Help. (n.d.). https://www.minitab.com/en-us/support/.

Box, G., & Draper, N. (1987). *Empirical model-building and response surfaces*. Wiley.

Cintas, P., Almagro, L., & Labres, X.-M. (2012). Industrial statistics with minitab.

George, M. L., Rowlands, D., Price, M., & Maxey, J. (2005). *Lean six sigma pocket Toolbook*. McGraw-Hill.

Gupta, H. C., Guttman, I., & Jayalath, K. P. (2020). *Statistics and probability with applications for engineers and scientists using MINITAB, R and JMP* (2nd ed.). Wiley.

Joiner, B. L. (1980). Lurking variables: Some examples. *The American Statistician*, 227–233.

Patterson, G., & Fedrico, M. (2006). Six sigma champions pocket guide. Rath & Strong.

Ryan, B., Joiner, B. L., & Cryer, J. (2012). *Minitab handbook*. Cengage.

Sleeper, A. (2012). *Minitab deMystified*. McGraw Hill.

Stagliano, A. A. (2004). *Six sigma advanced tools pocket guide*. McGraw Hill.

The Improve Phase

6

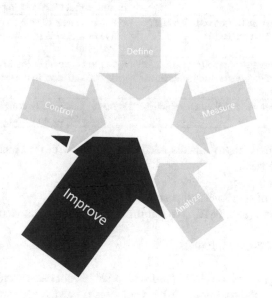

6.1 Introduction

Remember in the Improve phase we will focus on identifying solutions that align with the root causes identified in the Analyze phase while minimizing the risks of negative consequences. As with the prior phases, we should also ensure the outcomes of the Improve phase are included in a well-structured and easy to follow storyboard (Federico & Beaty, 2003; George et al., 2005; Patterson & Fedrico, 2006).

Typical Improve storyboard contents are as follows, which we will explore in more detail throughout this chapter:

© The Author(s), under exclusive license to Springer Nature Switzerland AG 2022
T. D. Blackburn, *Six Sigma*, https://doi.org/10.1007/978-3-030-96213-5_6

Table 6.1 KIND Karz, root causes and improvements needed

Root cause	Improvements needed
Test method is insufficient to detect pinhole leaks consistently	• Improve test method for pinhole leaks in airbags. • Ensure correct fabric is shipped.
Device B gives incorrect fabric thickness reading at the vendor	• Discontinue use of device B to measure fabric thickness.
High settings on temperature, vacuum, and dwell time impact seal strength	• Optimize settings on the new assembler machine (high settings on temperature, vacuum, and dwell time).
Fabric was exposed to excessive temperature in the warehouse—This was later determined after reviewing temperature data due to a power outage during a hurricane	• Store fabric in warehouse section served with a generator.
Incorrect fabric was shipped	Ensure correct fabric is shipped
Caliper used to measure torsion rod diameter doesn't have the proper resolution	• Purchase new torsion test device to ensure five or more distinct categories are available.
The deslagger cleaning device settings are causing the mold of the torsion rod to decrease in diameter over time	• Correct the caliper torsion mold cleaning device and reset to the original diameter.
The incorrect caliper was installed on vehicle 4 (pickup trucks)	• Install the properly designed calipers on vehicle 4 (pickup trucks).

- List of confirmed improvements needed (aligning with the Analyze phase confirmed root causes).
- Implementation plan.
- Cost estimate of solutions and cost/benefit analysis.
- Risk assessment (e.g., FMEA) to minimize the risk of negative consequences of new solutions.
- Sometimes—piloting solutions.

Begin the Improve phase storyboard with a list of root causes and improvements needed to address the question, "What root causes carried forward that I need to address?" These should carry forward from the Analyze phase. Here, there should be no surprises, and each should link to the prior phases and be easy to understand. Keep it simple at first—just list the root causes and then the activities needed to resolve them. Below is an example of that for KIND Karz (Table 6.1).

To ensure the root causes are resolved, effective project management tools and methods are needed. Specifically, Six Sigma projects use Implementation Plans to help facilitate the project management.

6.2   Implementation Plans

When developing an implementation plan, restate the improvements needed. Then, identify specific actions needed to ensure the root causes are resolved. In a structured way, this will address the question, "What are solutions that will resolve the root causes?"

As with any project management approach, list an owner for each action, due date, and status. For action item owners, this will be the specific person responsible for completing the activity. This answers the fundamental question, "Who will ensure the issues are resolved, and when are the issues needed to be resolved?" Also list the status which should be updated weekly (or every 2 weeks) in team meetings (where the status should be reviewed). At this point, the project team is taking on the role of a deployment team as well. For status, list whether the activity is *not started, on track, at risk, late, completed, or canceled*. Below are examples of Implementation Plans for KIND Karz (Tables 6.2 and 6.3).

Note that the Implementation Plans also include specific solutions. Next, we will review tools and methods to arrive at solutions necessary to resolve the root causes.

Table 6.2 KIND Karz, airbags, implementation plan

ID	Description	Action	Assigned to	By	Due date	Status
1A	Improve test method for pinhole leaks in airbags	Purchase new pinhole visualizer capable of detecting pinhole leaks	C. M. Martin	KIND Karz, QC	1-Mar-20	Not started
2A	Discontinue use of device B to measure fabric thickness	Require that the vendor only use device A going forward to measure fabric thickness	C. M. Martin	Vendor	1-Mar-20	Completed
3A	Optimize settings on the new assembler machine (high settings on temperature, vacuum, and dwell time)	Standardize settings on the new assembler to ensure high settings are applied for temperature, vacuum, and dwell time	L.W. Smythe	KIND Karz, Mfg ops	1-Mar-20	Late
4A	Store fabric in warehouse section served with a generator	Construct and insulate a section of the warehouse with sufficient size to store the fabric and provide a backup generator for HVAC	L.W. Smythe	KIND Karz, Mfg ops	1-Mar-20	At risk
5A	Ensure correct fabric is shipped	Require a procedure change to verify fabric at the vendor prior to shipment	C. M. Martin	Vendor	1-Mar-20	On track

Table 6.3 KIND Karz, brake calipers, implementation plan

ID	Description	Action	Assigned to	By	Due date	Status
1B	Purchase new torsion test device to ensure five or more distinct categories are available	Specify and purchase torsion test device that can measure at least five categories	C. M. Martin	KIND Karz, QC	1-Mar-20	Not started
2B	Correct the caliper torsion mold cleaning device and reset to the original diameter	Set the mold cleaning device to the default settings and ensure this is checked before each casting	C. M. Martin	KIND Karz, QC	1-Mar-20	Completed
3B	Install the properly designed calipers on vehicle 4 (pickup trucks)	Initiate recalls on all type 4 vehicle (pickup trucks) during the affected period and replace with correct caliper mechanisms	A. W. Gonzalez	KIND Karz, repair operations	1-Mar-20	Late

6.3   Arriving at Solutions

A danger as we begin to ideate solutions is the project team can return to bad habits,, perhaps even forgetting the work done so far in the project. It is important to link back to the prior work, especially the actions needed to resolve the root cause. It is important to remind the team of this as well.

There are a variety of tools and techniques that can be used depending on the nature of a given root cause needing to be resolved. Brainstorming, which was covered earlier in the book, is helpful to come up with ideas. Then tools such as the Effort-Impact matrix can help narrow down the ideas. This tool is shown in Table 6.4. Note that there are four boxes or cells in which to enter solutions. Top priority ideas will have a high impact but require low effort. Next are possible ideas, which have low impact but also low effort. The next box is associated with ideas that are high impact but also require high effort—these ideas need further development. Last, there are ideas that can usually be quickly ruled out, which have a low impact but require high effort (not a priority).

Below is an example for airbag solutions (Table 6.5) for KIND Karz. After the team brainstormed ideas, they settled on the following ideas. Note the solution for fabric being overheated needed further work to reduce the effort and cost (currently in high impact but also high effort). After brainstorming further, the KIND Karz project team found a way to isolate an area of the warehouse with prefabricated

Table 6.4 Effort-Impact matrix

	Low effort	High effort
High impact	Priority	Further develop the idea
Low impact	Possible	Not a priority

Table 6.5 KIND Karz, final Effort-Impact for airbag solutions

Airbag	Low Effort	High Effort
High Impact	3A, 2A, 5A, 1A	4A
Low Impact		

ID	Description
1A	Improve test method for pin hole leaks in air bags
2A	Discontinue use of Device B to measure fabric thickness
3A	Optimize settings on the new assembler machine (high settings on Temperature, Vacuum, and dwell Time)
4A	Store fabric in warehouse section served with a generator
5A	Ensure correct fabric is shipped

Table 6.6 KIND Karz, final Effort-Impact for caliber solutions

Brake	Low Effort	High Effort
High Impact	3B, 2B, 1B	
Low Impact		

ID	Description
1B	Purchase new torsion test device to ensure 5 or more distinct categories are available
2B	Correct the caliper torsion mold cleaning device, and reset to the original diameter
3B	Install the properly designed calipers on vehicle 4 (pick-up trucks)

walls that greatly reduced the cost and size of a generator needed. That would move 4A to the left.

Also shown below (Table 6.6) is the final Effort-Impact matrix for the brake caliper root cause resolution, which were all in the low effort, high impact category (which Albert was pleased to see).

So far we have covered key aspects of traditional project management, including schedule, scope, and who is responsible. But a key component remains—cost.

6.4   Cost-Benefit Analysis

Project management can be compared to a three-legged stool. The three legs are cost, schedule, and quality (or scope). If *any one* of the three are missing or not managed well, the project outcomes likely will not meet expectations. The same is true for Six Sigma projects, where cost should also be considered. For example, the team needs to ensure funding will be available. If not, consider whether the improvement idea be deferred, or look for creative alternative solutions. This leads to another question that needs to be answered, "Do I have sufficient funding, and a reasonable payback?"

Costs must be estimated or determined. In some cases, that is simple, such as purchasing a commercially off the shelf (COTS) item, but usually it is more complicated than that. In some cases, additional preliminary engineering design could be needed to facilitate a more accurate assessment of costs. Also historical cost databases can be helpful, comparing to similar projects.

Once the costs are estimated or identified, then further analysis is needed to ensure it represents a good investment for the company, or to compare alternatives where the improvement must be made (such as for safety and compliance or to meet the project objective). Engineering Economics offers a myriad of methods to arrive at a conclusion, but we will focus on two common methods—cost/benefit and payback methods and internal rate of return (IRR).

Cost/benefit and payback analysis is a simple concept that is easily explained and generally accepted. It looks at the number of years it will take to reclaim the investment. To illustrate how this works, let's return to the KIND Karz example. Remember there are problems with the airbag fabric becoming separated from the housing. To prevent this, the fabric vendor has proposed a new end seam and improved adhesive at the same cost, but it requires a new assembler head. This will need to be charged back to KIND Karz as it will only be used for their airbag type. The assembler vendor estimates it will cost $25,000 per assembler, and there are three assemblers (for a total of $75,000). However, the current costs of rejects are estimated to be $50,000 per year, and the vendor believes this solution will reduce defects by 90%. This is a challenging situation—lives are at stake, as well as reputation. However, if the cost of this does not meet with KIND Karz financial objectives, another solution will need to be considered.

Companies typically have different year payback targets depending on the investment types. For technologies such as this, KIND Karz usually requires a 3-year payback given how frequently technologies change. Once that is known, the calculation is simple—just divide the total cost by the expected annual savings:

- Total cost: $75,000.
- Annual savings: $0.9 \times \$50,000 = \$45,000$.
- Therefore, years to payback = $\$75,000/\$45,000 = 1.7$ years.
- Since the payback is less than 3 years, the improvement idea is accepted.

Other companies might use an internal rate of return (IRR) method, which determines the rate of return that will result in the net present value (NPV) equal to zero, which is compared to a hurdle rate, or desired rate of return. This is a well-established method and has been used for many years (Dorfman, 1981). If the IRR is greater than the hurdle rate, the project is considered financially viable. IRR can also be used to compare alternatives. The following is the equation:

$$0 = NPV = \sum_{t=0}^{T} \frac{C_t}{(1-IRR)^t}$$

where

- C_t = Net cashflow during the period t
- IRR = the internal rate of return.
- t = the number of time periods.

Fortunately, Excel can easily be used to calculate this, using the =IRR () equation. As shown below in Fig. 6.1, note the initial row is the starting point, which includes a cost of $75,000, followed by the annual savings of $45,000. The IRR is calculated to be 35.31%. KIND Karz has a hurdle rate of 8%, so this would be considered a satisfactory investment.

But what if Albert discovered at the end of the 3 years that there will be a $10,000 removal and disposal cost? Would this option still meet the hurdle rate? The same method can be used but reduce the third year savings by $10,000 (see Fig. 6.2). The IRR is still greater than the hurdle rate, so this remains a viable option.

Fig. 6.1 Excel IRR calculation

=IRR(C2:C5)

Year	Cash Flow
0	$(75,000)
1	$ 45,000
2	$ 45,000
3	$ 45,000
IRR:	36.31%

Fig. 6.2 IRR with disposal included

=45000-10000

Year	Cash Flow
0	$(75,000)
1	$ 45,000
2	$ 45,000
3	$ 35,000
IRR:	32.14%

Another question regarding costs needs to be considered, which is "If there are more solutions than time or funding available, how many do I need to meet my project goal or objective?" When this occurs, it is helpful to prioritize which solutions can be done and perform a pro forma estimate of the total improvement to confirm it aligns with the project goals. For KIND Karz, after meeting with the CFO (Chris) as to the costs of improvements, Albert was authorized to proceed.

So far the, the KIND Karz team has identified improvements, determined the timing, assigned owners, and confirmed the costs are feasible. However, it is also important to ensure we minimize the risk of the improvements resulting in unwanted effects. This is related to the remaining question, "What are the risks of the solutions, and how will any unacceptable risks be abated?"

6.5   Risk Analysis

To ensure the improvements won't result in unwanted effects, a risk analysis is needed. A common tool used is the FMEA, or the Failure Mode and Effect Analysis tool. In the Improve phase, create a table that lists all the proposed improvements. Add next to it a Failure Mode column. In this column, list what could go wrong because of the improvement. There is often more than one event that could occur. Then add another column, which is the Failure Effects column. Here enter the consequences of the Failure Mode.

Finally, create the risk scores. The FMEA has three scores—Severity, Likelihood of Occurrence, and Likelihood of Detection. It is common to score 1–10 for each of these, with 1 being the least and 10 being the most severe. After all three scores are entered, then the product of all three represent the Risk Priority Number.

See Table 6.7 for the KIND Karz Airbag improvement ideas. For example, note solution 2A, which will direct the vendor to only use Device A going forward to measure the fabric for purposes of calibrating the machine. Note that the team identified two failure modes—the correct device could also deteriorate over time, or the vendor could revert to the flawed device. In either case, the Severity (SEV) score will be high (9), as we know the thin fabric has already led to airbag failures. However, the team thinks it is only a moderate chance (5) of Device A deteriorating and a low chance (1) the vendor will revert to the prior device. Both have a moderate (5) chance of being detected. The resulting Risk Priority Numbers (RPN) are 225 and 45. Therefore, the first of the two failure modes are of a greater risk.

However, we don't want to stop here. Highest RPNs should receive greater diligence to reduce the risks, but as a practical matter seek to mitigate risks of any unacceptable risk (e.g., high SEV scores), or solutions with RPNs greater than 100.

Albert reassembled the team and explained that several of the solutions offer unacceptable risks. The team brainstormed further, and came up with risk mitigating ideas, and rescored the risks. For the prior example for the 2A solution, the team came up with the idea for the vendor to compare Device A to a known standard monthly to ensure it would not drift over time without their being aware of it. This reduced the Likelihood of Detection score to 1, and the resulting RPN reduced from 225 to 45. The vendor agreed to do this, so the risk was abated (see Table 6.8).

Table 6.7 FMEA, KIND Karz airbag solutions

ID	Action	Failure Mode	Failure Effects	SEV	OCC	DET	RPN
1A	Purchase new pinhole visualizer capable of detecting pinhole leaks	Miss detecting the pinholes	Airbag tear	9	5	9	405
2A	Require that the vendor only use Device A going forward to measure fabric thickness	Device A also deteriorates over time	Too thin fabric	9	5	5	225
	Ditto	Reverts to Device B	Too thin fabric	9	1	5	45
3A	Standardize settings on the new assembler to ensure high settings are applied for Temperature, Vacuum and Dwell Time	Revert to other assembler settings	Airbag tear	9	5	9	405
4A	Construct and insulate a section of the warehouse with sufficient size to store the fabric and provide a back-up generator for HVAC	Generator does not start when power is out	Airbag adhesion failure	5	5	1	25
5A	Require a procedure change to verify fabric at the vendor prior to shipment	Failure to follow procedure	Incorrect fabric, Airbag tear	9	1	9	81

When using the FMEA tool, there are few tips and watchouts. First, when recalculating RPN after identifying risk reduction measures, many often make the mistake of reducing SEV. However, this rarely should be reduced. Remember SEV is the score of Impact if the event were to occur and is not impacted by likelihood or detection ability. For another tip, it's often better to populate most of the FMEA before reviewing it with your team, although have discussions with SMEs on what could go wrong informally prior to completing it. When scoring, it is easy to get into a team debate. To avoid this, agree ahead of time to use 1, 5, 9. Finally, remember to add the risk reduction steps to the implementation plan, or carry over to the Control phase.

Table 6.8 FMEA, KIND Karz, airbag solution risk mitigation

ID	Action	Failure Mode	Risk Reduction Measures	SEV	OCC	DET	RPN
1A	Purchase new pinhole visualizer capable of detecting pinhole leaks	Miss pinholes	Repeat Attribute Agreement and adjust as needed	9	1	1	9
2A	Require that the vendor only use Device A going forward to measure fabric thickness	Device A also deteriorates over time	Compare Device A to a standard monthly	9	5	1	45
	Ditto	Reverts to Device B	No action required	9	1	5	45
3A	Standardize settings on the new assembler to ensure high settings are applied for Temperature, Vacuum and Dwell Time	Revert to other assembler settings	Add settings to standard set-up guides and required verification signature	9	1	5	45
4A	Construct and insulate a section of the warehouse with sufficient size to store the fabric and provide a back-up generator for HVAC	Generator does not start when power is out	No action required	5	5	1	25
5A	Require a procedure change to verify fabric at the vendor prior to shipment	Failure to follow procedure	Verifying signature in shipping document; enhance	9	1	1	9
ID	Action	Failure Mode	Risk Reduction Measures	SEV	OCC	DET	RPN
			receipt testing at KIND Karz				

The FMEA can also be useful in other ways as well. The following are other uses:

- To help identify factors to consider in the Measure phase.
- When designing new processes.
- Improving an existing process.
- Identify potential root causes.
- When applying things designed for something else.
- When applying designs to another application.

Now that we have reviewed the risks and developed risk mitigating ideas, there might be some of the solutions that could be further evaluated and improved by piloting.

6.6 Piloting

As noted in a prior chapter, piloting can occur in the Analyze phase, but also it is common in the Improve phase. Whereas it is used in the Analyze phase to provide evidence as to root cause, in the Improve phase it is useful to identify real-life issues or failures that might occur and enable the team to resolve them before full deployment. Usually, pilots are small scale with a limited scope or cost. Ultimately, piloting can make the full implementation more effective and efficient by exposing issues and risks that might not be anticipated.

Design the pilot carefully, and assume you only get one chance at it. Also complete an FMEA for other features of the pilot before proceeding. Work with your team to establish clear objectives. Also establish clear procedures and expectations. And if you are the leader, be there and remain engaged.

Continue the pilot long enough to gather meaningful data. As early wins emerge, share them, and celebrate to get further buy-in, and use early winnings to recognize the team. Be open minded and learn from the good and the bad. Also be willing to accept defeat if needed and be flexible to look for an alternative solution or better approach.

Once the pilots are complete, quickly move to implementation. Once the implementations are complete, you are not finished—make sure the root causes were resolved and the project goals were met. Also make sure the issues do not reoccur. These occur in the Control phase.

The following are some references related to this chapter.

References

Dorfman, R. (1981). The meaning of internal rates of return. *The Journal of Finance*, 1011–1021.

Federico, M., & Beaty, R. (2003). *Six sigma team pocket guide*. McGraw-Hill.

George, M. L., Rowlands, D., Price, M., & Maxey, J. (2005). *Lean six sigma pocket Toolbook*. McGraw-Hill.

Patterson, G., & Fedrico, M. (2006). *Six sigma champions pocket guide*. Rath & Strong.

The Control Phase

7.1   Introduction

Remember in the Control phase we will confirm the project met its objective (with statistical significance). We will ensure means are established to ensure the problems won't reoccur. As with prior phases, document the Control phase in the storyboard. Typical Control storyboard contents are as follows, which we will explore in more detail throughout this chapter:

- Confirm objectives were achieved.
- Summary of objectives achieved.
  - Data and process analysis for verification.
  - Monitoring and control strategy.
- Standardization.
- Hand-offs.
- Key learnings.
- Carryover projects or initiatives.

T. D. Blackburn, *Six Sigma*, https://doi.org/10.1007/978-3-030-96213-5_7

7.2   Confirming Objectives Were Achieved

The first question we need to answer is, "Did I meet my objective? If so, was the improvement statistically significant?" To do so, statistical tools are needed, which typically are the same ones introduced earlier in this book. For example, a variety of tests can be used to show significance, such as Two Sample T Test, ANOVA, test of proportions, and others. Also control charts are helpful to confirm the process is stable, and process capability analysis is used to confirm the process can achieve the desired specifications (George et al., 2005; Ryan et al., 2012; Stagliano, 2004; Sleeper, 2012; Ryan et al., 2012).

To illustrate this, consider the KIND Karz case study. Remember from the Define phase the current baseline was 200 warrantee or recall claims per 1000 automobiles sold after 12 months. To confirm our project met its primary objective, we need to confirm we have met the goal of no more than 100 recalls or warrantee claims per 1000 automobiles sold. We also had a stretch goal of no more than 50 per 1000.

The team was excited to learn, after the improvements were put in place, whether the goal was met. Albert shared the before and after data as shown in Table 7.1. Note

Table 7.1   KIND Karz before and after performance

Month	Cars produced	Warrantees and recalls	Timing
Sep 2019	1007	195	Before
Oct 2019	1008	209	Before
Nov 2019	964	200	Before
Dec 2019	1000	200	Before
Jan 2020	1016	195	Before
Feb 2020	1017	199	Before
Mar 2020	991	196	Before
Apr 2020	1008	206	Before
May 2020	1020	195	Before
Jun 2020	968	206	Before
Jul 2020	970	197	Before
Aug 2020	971	207	Before
Sep 2020	1014	209	Before
Oct 2020	954	205	Before
Nov 2020	1007	195	Before
Dec 2020	1008	209	Before
Apr-21	964	45	After
May-21	1000	52	After

(continued)

Table 7.1   (continued)

Month	Cars produced	Warrantees and recalls	Timing
Jun-21	1016	51	After
Jul-21	1017	49	After
Aug-21	991	52	After
Sep-21	1008	50	After
Oct-21	1020	52	After
Nov-21	991	45	After
Dec-21	1008	49	After
Jan-22	1020	52	After
Feb-22	968	49	After
Mar-22	970	45	After
Apr-22	971	51	After

that the number of cars produced is shown and the warrantees or recalls per 1000. Another column is added to show the timing of the data.

First, Albert created a control chart, as shown in Fig. 7.1. Since the data is proportional (warrantees and recalls/cars produced), he created a P-chart. Note the prior proportion was about 20% but is now 5%. This would suggest the goal and the stretch goal was achieved. However, is this change statistically significant?

To confirm whether the improvement was statistically significant, Albert proceeded to run a Two Sample T Test. The box plot (Fig. 7.2) certainly suggests there is a difference and an improvement, and the Minitab output (Fig. 7.3) confirms it. Note the P-value is 0.000 which is less than alpha = 0.05, so we can reject the null hypothesis and accept the alternative hypothesis that there is a statistically significant improvement.

Next, Albert rechecked process capability using the discrete method. Recall the prior Ppk was 0.28 (Bothe, 1997). He reported these as estimates since the data after did not have at least 100 values and will continue to calculate and report until 100 samples are acquired (Figs. 7.4 and 7.5).

The next step is to summarize the achievement of the original objective. In a storyboard, the summary table shown in Table 7.2 is usually provided, followed by the statistical analysis. Note the table shows the same criteria as the CTQ in the Charter (all the way back to the Define phase) and matches the goal or objective. The table also includes the results and the method of verification.

That was good news for the team. But Albert reminded them, "Well done! But we're not done yet. We need to answer other questions, such as how will we ensure the improvements are sustained? How will be know if we start to drift back to the earlier recall and warrantee claim numbers?"

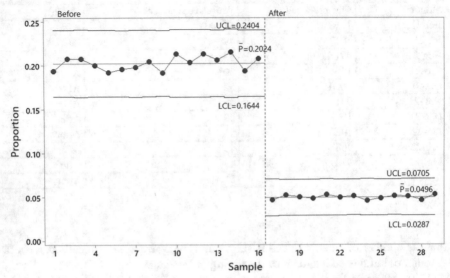

Fig. 7.1 KIND Karz control chart, before and after

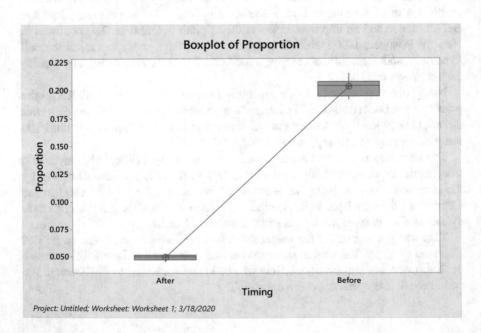

Fig. 7.2 KIND Karz before/after improvements chart

Two-Sample T-Test and CI: Proportion, Timing

Method

μ_1: mean of Proportion when Timing = After
μ_2: mean of Proportion when Timing = Before
Difference: $\mu_1 - \mu_2$

Equal variances are not assumed for this analysis.

Descriptive Statistics: Proportion

Timing	N	Mean	StDev	SE Mean
After	13	0.04959	0.00237	0.00066
Before	16	0.20253	0.00789	0.0020

Estimation for Difference

Difference	95% CI for Difference
-0.15294	(-0.15731, -0.14857)

Test

Null hypothesis	$H_0: \mu_1 - \mu_2 = 0$
Alternative hypothesis	$H_1: \mu_1 - \mu_2 \neq 0$

T-Value	DF	P-Value
-73.56	18	0.000

Boxplot of Proportion

Fig. 7.3 KIND Karz before/after Two Sample T Test outputs

Category	Entry or Results
Proportions Defective	20.276%
Z	0.8318
Ppk	0.28

Fig. 7.4 Prior Ppk for KIND Karz

Category	Entry or Results
Proportions defective	5%
Z	1.645
Ppk	0.55

Fig. 7.5 Post-improvement Ppk, KIND Karz

Table 7.2 KIND Karz summary of objectives achieved

Criteria	Objective	Results	Method of verification
Total recalls/ warrantee work	No more than 100/1000 cars	50/1000 cars	P-chart, test of two proportions (P-value = 0 < alpha 0.05)

7.3   Monitoring and Control Strategy

It is important to ensure that over time wins are sustained. To do so, the process needs to be properly monitored. Control charts are commonly used to detect shifts and other special cause variation, but other means might be needed depending on the specific application. First focus on the big Y or the overall CTQ (for the KIND Karz example, that was the proportion of recalls and warrantees in the first 12 months). However, it is also important to monitor the little Ys as well. In the KIND Karz case study, that was for the airbag and brake caliper issues. These should at least be monitored in the months following the project to ensure no drift or returning to old approaches occurs.

To monitor the Y's, it is helpful to develop a table. An example for KIND Karz is shown in Table 7.3. This helps to answer the question, "How will we know if we start to drift?" Note that the description of what will be monitored is shown first. It includes the big and small Ys. Then a monitoring strategy is identified. Since for this example all three include failure rates, the P-chart will be used. The data source is identified, as well as who is responsible for collecting and analyzing the data. It also identifies the *who* will review the data, as well as whom it will be escalated to when there are special cause concerns.

Often, additional monitoring is needed for improvements put in place, at least for a period until there is confidence the new ways of working have become common. The tables are similar, and examples from KIND Karz for the airbag and brake improvements are shown in Tables 7.4 and 7.5.

In part, identifying monitoring and control strategies helps also answer the question, "How will we ensure the improvements are sustained?" However, standardization is needed in addition to monitoring to ensure there is consistency in practices over time.

Table 7.3   KIND Karz monitoring and control strategy

Description	Monitoring strategy	Data source	Tracked by	Reviewed	Escalation
Warrantee and recall rates after the first 12 months of vehicle sale	Rolling 12-month P-chart	Recalls department	OpEx	Monthly management quality reviews	To VP of quality for special cause events
Airbag failure rates	Rolling 12-month P-chart	Service department	OpEx	Monthly management quality reviews	To VP of quality for special cause events
Brake caliper failure rates	Rolling 12-month P-chart	Service department	OpEx	Monthly management quality reviews	To VP of quality for special cause events

Table 7.4 KIND Karz airbag improvements monitoring and control strategy

ID	Action	Control strategy	Control resp	Monitoring strategy	Monitoring resp	Monitoring frequency
1A	Purchase new pinhole visualizer capable of detecting pinhole leaks	Perform attribute agreement analysis on any new analysts	OpEx department	Annual quality reviews	Audits department	Annual
2A	Require that the vendor only use device A going forward to measure fabric thickness	Confirm device A us used as part of biannual vendor audits	Audits	Check material type and thickness on receipt also	Receiving	Annual
3A	Standardize settings on the new assembler to ensure high settings are applied for temperature, vacuum, and dwell time	Add settings to standard setup guides and required verification signature	Operations	Control chart for defect rates	Operations	By shift
5A	Require a procedure change to verify fabric at the vendor prior to shipment	Verifying signature in shipping document; enhance receipt testing at KIND Karz	Recalls department	Confirm signatures complete on receipt	Receiving	Monthly

Table 7.5 KIND Karz brake caliper improvements monitoring and control strategy

ID	Action	Control strategy	Control resp	Monitoring strategy	Monitoring resp	Monitoring frequency
1B	Specify and purchase torsion test device that can measure at least five categories	Calibrate against a known standard once a quarter	Calibrations department	Review as part internal audits	Audits department	Annual
2B	Set the mold cleaning device to the default settings and ensure this is checked before each casting	Calibrate against a known standard once a quarter	Operations	Control chart for rod diameter	Audits department	Annual
3B	Initiate recalls on all type 4 vehicle (pickup trucks) during the affected period and replace with correct caliper mechanisms	Offer an incentive to return for repair and send notices to state inspection stations	Recalls department	Track percent complete and escalate contact approach over time	Recalls department	Monthly

7.4   Standardization

In addition to monitoring and control, work processes need to be standardized to ensure improvements are sustained. Standardization can occur in many forms, such as standard operating procedures, setup guides, preventive maintenance, signage, training and development, forms, signage, and software systems (new or updates such as signals). This also answers the question, "How do we then standardize new ways of working?"

It is helpful to create a table as to how standardization will occur. For the KIND Karz example, see in Tables 7.6 and 7.7. Note there is an action or improvement that was implemented, and then the team identified ways to standardize in the last column.

Also important is the question, "How do we pass along any new ways of working to employees and management?" This of course requires training on things that have changed and new processes. However, this might sometimes be the only form of standardization (or even the corrective action). But while training usually is not the primary solution or standardization method, it is frequently needed to accompany other more impactful changes. When designing training, ensure it focuses on what changed and includes existing operators. Typically, auto sequences (things we

Table 7.6 KIND Karz standardization, airbags

ID	Action	Standardization
1A	Purchase new pinhole visualizer capable of detecting pinhole leaks	Revise pinhole test procedure and form and retrain analysts
2A	Require that the vendor only use device A going forward to measure fabric thickness	Ensure the vendor changes their procedure. Also change the quality agreement with the vendor to reflect this.
3A	Standardize settings on the new assembler to ensure high settings are applied for temperature, vacuum, and dwell time	Change the setup procedure and retrain setup technicians. Add signage for correct settings
5A	Require a procedure change to verify fabric at the vendor prior to shipment	Ensure the vendor changes their procedure. Also change the quality agreement with the vendor to reflect this. Also change receiving procedure and retrain KIND Karz employees

Table 7.7 KIND Karz standardization, brake calipers

Action	Standardization
Specify and purchase torsion test device that can measure at least five categories	Revise test procedure to reflect correct device. Train testers. Change calibration procedure
Set the mold cleaning device to the default settings and ensure this is checked before each casting	Modify the settings in the setup guide. Train set-up technicians
Initiate recalls on all type 4 vehicle (pickup trucks) during the affected period and replace with correct caliper mechanisms	N/A

do from habit) are well embedded in operators ways of working, and it can be difficult to train on a change. There is a risk they will forget and revert to old ways of working. It is helpful to put signals where changes occur, and allow time to practice, reinforcing the new auto sequences.

As noted, some may over rely on training or retraining to resolve a root cause. Only when a root cause was for the lack of knowledge or skills would such retraining be helpful usually. Also, if the person was trained previously and it was ineffective, just retraining without correcting the design of the learning content and approach should not be expected to result in improvement. Also, retraining isn't helpful when people forget—signals must be added (in documents or visual/audible in the workplace) to prevent the mistake from happing again. Finally, try to build in error-proof measures to prevent the possibility that a human error could occur when possible.

7.5 Project Closure and Hand-off

So far, this book has illustrated how to resolve all the questions introduced in Chap. 1, with exception to three remaining questions: "To whom and how do we hand off responsibilities to own this going forward? What did we learn from this project? Are there any off-shoot projects, or *just do it* activities that need to follow?"

First, hand off the project. Be aware others might expect you to own the project forever. While you certainly should be expected to answer questions later and do some follow-ups in the coming months, the overall responsibility should return to the project owner. Before closing the project, secure consent from the project owner and sponsor that it is ready to close. Make sure the monitoring and control program is implemented and results recorded and reported as defined prior. Identify and communicate longer-term actions identified but not possible for the project to resolve.

Likely you will need to continue to have some level of involvement for a few weeks or months and assist resolving any surprising effects of the project. While each project is different, it is helpful to check in at least monthly for the 3 months following the project closure to ensure practices are effective and in place. Then, have a formal check-in 6 months and 1 year following the project close.

Also consider lessons learned and include those in the storyboard. For example, consider if there are any insights that can be applied in similar situations (e.g., machine centers, similar lines, products, etc.). Also learn from the project process and consider if anything should have been done differently. Consider how to preserve the learnings even outside the project and how future issues can be avoided. Usually, spinoff projects can also be identified and assigned.

The team considered what they had learned in the KIND Karz project. Many were skeptical at first, but all had come to believe in the DMAIC, and data-driven approach Albert had patiently insisted on. Some of the lessons learned were as follow. Of course, these were specific to the KIND Karz project—yours will be different.

- Don't assume testing devices and instruments are reliable.
- Need for comprehensive quality agreements and increased quality oversight to vendors.
- Importance of maintenance and setup.
- Ensuring new equipment is evaluated for its own operating criteria.

Next, assemble documents and store them appropriately (especially electronically). Clean up and store the storyboard in a location that is accessible to appropriate individuals. Then don't forget to recognize the team and celebrate successes. Usually, a simple *thank you* goes a long way, as well as recognition in town halls, communications, and other events.

The following are some references related to the Control phase.

References

Cintas, P., Almagro, L., & Labres, X.-M. (2012). Industrial statistics with Minitab.

Federico, M., & Beaty, R. (2003). *Six sigma team pocket guide*. McGraw-Hill.

George, M. L., Rowlands, D., Price, M., & Maxey, J. (2005). *Lean six sigma pocket Toolbook*. McGraw-Hill.

Minitab Online Help. (n.d.). https://www.minitab.com/en-us/support/.

Patterson, G., & Fedrico, M. (2006). *Six sigma champions pocket guide*. Rath & Strong.

Ryan, B., Joiner, B. L., & Cryer, J. (2012). *Minitab handbook*. Cengage.

Sleeper, A. (2012). *Minitab DeMystified*. McGraw Hill.

Stagliano, A. A. (2004). *Six sigma advanced tools pocket guide*. McGraw Hill.

Storyboards

8

Project storyboards are recommended for Six Sigma projects. Presentation software such as Microsoft © PowerPoint are helpful, as the same material can be used for presentations as well as archival. But ensure the storyboard is clear and easy to understand. As a general good practice, don't make the reader *work at it* to understand the content. Don't write to impress, and tailor the findings in such that they are comprehensible to a reasonably informed audience. There should be a clear story and thread throughout the storyboard, with one phase linking to another clearly.

Each graph or figure should include a text box with why and an interpretation. Follow good presentation writing skills, such as limit contents five to eight bullets. Don't use fancy font or backgrounds. Also write it in such that you can quickly recall the content, even after a period of time. As a rule of thumb, write the storyboards such that you can come back to it in 5 years and quickly recall and explain.

Craft the storyboards by DMAIC phase, continuing to add to it as additional information becomes available. Avoid the temptation to have multiple storyboards for different audiences—that will be difficult to maintain and is completely unnecessary. Instead, use the hide feature to tailor a presentation for a given audience.

The following are recommended minimum contents of storyboards.

8.1 Define Phase Storyboard Recommended Contents

- Project charter
- CTQ hierarchy
- SIPOC diagram
- Process map
- Stakeholder analysis (remember not to include the stakeholder map or anything potentially inflammatory, however, in the storyboard but keep separate)
- Learnings from similar experiences
- Background information as needed

T. D. Blackburn, *Six Sigma*, https://doi.org/10.1007/978-3-030-96213-5_8

8.2   Measure Phase Storyboard Recommended Contents

- Additional process maps if needed
- Potential predictors (X's)—Factor Priority Matrix
- Data collection plan
- Measurement system analysis (MSA)
- Process stability
- Process capability
- Stratification
- Summary of the Measure phase

8.3   Analyze Phase Storyboard Recommended Contents

- Questions and scope to carry forward from the Measure phase
- Identification of potential root causes
- Root cause analysis visuals (fishbone or Causal Tree)
- Accept or reject based on evidence
    ◦   Process analysis
    ◦   Data analysis
    ◦   In-scope analysis
- Possible: Outcomes of pilots or DOEs
- Summary of the Analyze phase (verified and in-scope root causes)

8.4   Improve Phase Storyboard Recommended Contents

- List of confirmed improvements needed (aligning with the Analyze phase confirmed root causes)
- Implementation plan
- Cost estimate of solutions and cost/benefit analysis
- Risk assessment (e.g., FMEA) to minimize the risk of negative consequences of new solutions
- Sometimes—piloting solutions

8.5   Control Phase Storyboard Recommended Contents

- Confirm objectives were achieved
- Summary of objectives achieved
- Data and process analysis for verification
- Monitoring and control strategy
- Standardization
- Hand-offs
- Key learnings
- Carryover projects or initiatives

Appendixes

Appendix A: Introduction to Minitab V19–21

This Appendix is intended to be a quick familiarization for a new user to Minitab and is based on Minitab V20, which is essentially the same as V19 and V21 in appearance and navigation. Note that V18 and earlier will appear somewhat different. However, most functionality included in this book are similar between recent versions. (Most of the images in this book are based on V18 given the shaded background from that version provided better resolution, but functionality is provided where primarily for V19–21).

Minitab is an intuitive application, and this Appendix is meant to get the user started. Try performing some analysis with simple datasets or follow any of the steps in job aids found in this book for quick familiarity. You will get more comfortable as you use it.

When opening Minitab, the following views appear as described below and as illustrated in the graphic Fig. A.1. Note that the views can be changed by selecting "View."

- **Output pane** is where analysis information is provided (like *Session Window* in version 18 and earlier).
- **Command line** allows direct entry of commands.
- **Navigator** organizes and allows for quick access to various prior analysis and charts.
- **Data** view has a similar appearance to a worksheet and is where source data is contained. Data can be copied from another source such as Excel and pasted here.
- **History** pane.

Next notice the ribbon at the top, like Windows applications. The two most frequently used menu items are "Stat" and "Graphs." Most statistical analysis tools are found under "Stats." When selecting a tool, a dialogue box will appear, similar as shown in the job aids in this book. Stand-alone graphs are under the "Graphs" menu item (other graphs are associated with statistical tools).

© The Author(s), under exclusive license to Springer Nature Switzerland AG 2022
T. D. Blackburn, *Six Sigma*, https://doi.org/10.1007/978-3-030-96213-5

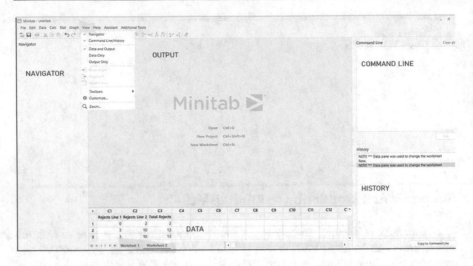

Fig. A.1 Minitab 20 layout

Files can be opened and saved under "File." In addition, go to "Options" under "File" to change defaults. This is helpful upon installation to set preferences or for company-specific changes.

The "Data" menu also contains helpful features. These include tools to rearrange or better organize data. Beside it is the "Calc" menu, where the Minitab calculator is included (can be used to create new columns of derived data using arithmetic operations and other columns). Also, it has a helpful random number generator, sorting, and other helpful features.

Appendix B: Datasets

While the datasets used in the examples for this book are available electronically. It is also included in this Appendix.

ANOVA dataset

Lot 1 failed calipers	Lot 2	Lot 3	Lot 4
409.5936511	412.598	413.5951	416.909
406.0177488	424.4588	417.8856	404.4245
439.9258497	419.445	423.2698	418.4069
391.4056183	398.3861	420.4561	395.683
403.6610695	408.1994	409.5607	419.9508
421.5128446	418.5715	423.7854	423.0175
421.2601833	418.3844	431.3619	413.9323
384.6860169	403.021	409.5141	408.6852
390.0692373	413.2736	402.7306	419.0362
417.0507987	425.8978	408.325	429.5754
409.4893464	435.0424	394.182	420.3918

Lot 1 failed calipers	Lot 2	Lot 3	Lot 4
426.584	420.5148	410.897	403.9304
401.4135369	419.3225	414.5562	414.7995
408.1949713	424.5787	408.3749	407.5027
421.2156261	419.4411	409.5534	413.6285
428.0041026	396.9029	406.1945	421.4388
422.4803311	437.3384	430.8973	421.4288
417.3703614	416.7981	407.9135	415.2077
413.8084468	420.8438	418.0728	405.7014
420.3436323	407.0863	419.978	408.4529
410.8339961	416.6585	427.2569	402.7221
422.5741033	423.5999	426.8836	408.1393
423.5797427	437.3816	407.6381	421.9519
400.0169238	407.4264	428.3304	401.8379
406.4848367	429.7692	410.565	411.9284
418.5894638	413.5771	403.8188	399.5491
446.0477413	414.0303	420.0103	403.4756
400.6996305	429.525	408.7593	415.1773
426.1775047	402.2992	408.7743	418.7407
405.9597955	420.2617	421.1944	427.9788

Attribute agreement dataset

ID	Appraiser	Response	Sample	Trial	Standard
1	John	5	1	1	5
2	John	5	1	2	5
3	John	3	2	1	3
4	John	3	2	2	3
5	John	1	3	1	1
6	John	1	3	2	1
7	John	2	4	1	2
8	John	2	4	2	2
9	John	1	5	1	1
10	John	1	5	2	1
11	John	2	6	1	3
12	John	3	6	2	3
13	John	4	7	1	4
14	John	4	7	2	4
15	John	5	8	1	5
16	John	5	8	2	5
17	John	2	9	1	2
18	John	2	9	2	2
19	John	1	10	1	1
20	John	1	10	2	1
21	John	5	11	1	5
22	John	5	11	2	5
23	John	4	12	1	4

ID	Appraiser	Response	Sample	Trial	Standard
24	John	4	12	2	4
25	John	1	13	1	1
26	John	1	13	2	1
27	John	3	14	1	3
28	John	3	14	2	3
29	John	2	15	1	2
30	John	2	15	2	2
31	John	4	16	1	4
32	John	4	16	2	4
33	John	1	17	1	1
34	John	1	17	2	1
35	John	2	18	1	2
36	John	2	18	2	2
37	John	3	19	1	3
38	John	3	19	2	3
39	John	5	20	1	5
40	John	5	20	2	5
41	John	5	21	1	5
42	John	4	21	2	5
43	John	4	22	1	4
44	John	4	22	2	4
45	John	3	23	1	3
46	John	3	23	2	3
47	John	2	24	1	2
48	John	2	24	2	2
49	John	2	25	1	2
50	John	2	25	2	2
51	John	4	26	1	4
52	John	4	26	2	4
53	John	1	27	1	1
54	John	1	27	2	1
55	John	2	28	1	2
56	John	2	28	2	2
57	John	5	29	1	5
58	John	5	29	2	5
59	John	3	30	1	3
60	John	3	30	2	3
61	John	5	31	1	5
62	John	5	31	2	5
63	John	2	32	1	2
64	John	2	32	2	2
65	John	1	33	1	1
66	John	1	33	2	1
67	John	3	34	1	3
68	John	3	34	2	3
69	John	1	35	1	1

ID	Appraiser	Response	Sample	Trial	Standard
70	John	1	35	2	1
71	John	3	36	1	2
72	John	2	36	2	2
73	John	4	37	1	5
74	John	5	37	2	5
75	John	4	38	1	4
76	John	4	38	2	4
77	John	2	39	1	2
78	John	2	39	2	2
79	John	3	40	1	3
80	John	3	40	2	3
81	John	1	41	1	1
82	John	1	41	2	1
83	John	2	42	1	2
84	John	2	42	2	2
85	John	2	43	1	2
86	John	3	43	2	2
87	John	3	44	1	3
88	John	3	44	2	3
89	John	3	45	1	3
90	John	3	45	2	3
91	John	5	46	1	5
92	John	5	46	2	5
93	John	4	47	1	4
94	John	4	47	2	4
95	John	5	48	1	5
96	John	5	48	2	5
97	John	4	49	1	4
98	John	4	49	2	4
99	John	1	50	1	1
100	John	1	50	2	1
101	Sheila	5	1	1	5
102	Sheila	5	1	2	5
103	Sheila	3	2	1	3
104	Sheila	3	2	2	3
105	Sheila	1	3	1	1
106	Sheila	1	3	2	1
107	Sheila	2	4	1	2
108	Sheila	2	4	2	2
109	Sheila	1	5	1	1
110	Sheila	1	5	2	1
111	Sheila	2	6	1	3
112	Sheila	2	6	2	3
113	Sheila	4	7	1	4
114	Sheila	4	7	2	4
115	Sheila	5	8	1	5

ID	Appraiser	Response	Sample	Trial	Standard
116	Sheila	5	8	2	5
117	Sheila	2	9	1	2
118	Sheila	2	9	2	2
119	Sheila	1	10	1	1
120	Sheila	1	10	2	1
121	Sheila	5	11	1	5
122	Sheila	5	11	2	5
123	Sheila	4	12	1	4
124	Sheila	4	12	2	4
125	Sheila	1	13	1	1
126	Sheila	1	13	2	1
127	Sheila	3	14	1	3
128	Sheila	3	14	2	3
129	Sheila	2	15	1	2
130	Sheila	2	15	2	2
131	Sheila	4	16	1	4
132	Sheila	4	16	2	4
133	Sheila	1	17	1	1
134	Sheila	1	17	2	1
135	Sheila	2	18	1	2
136	Sheila	2	18	2	2
137	Sheila	3	19	1	3
138	Sheila	3	19	2	3
139	Sheila	5	20	1	5
140	Sheila	5	20	2	5
141	Sheila	5	21	1	5
142	Sheila	5	21	2	5
143	Sheila	4	22	1	4
144	Sheila	4	22	2	4
145	Sheila	3	23	1	3
146	Sheila	3	23	2	3
147	Sheila	2	24	1	2
148	Sheila	2	24	2	2
149	Sheila	2	25	1	2
150	Sheila	2	25	2	2
151	Sheila	4	26	1	4
152	Sheila	4	26	2	4
153	Sheila	1	27	1	1
154	Sheila	1	27	2	1
155	Sheila	2	28	1	2
156	Sheila	2	28	2	2
157	Sheila	5	29	1	5
158	Sheila	5	29	2	5
159	Sheila	3	30	1	3
160	Sheila	3	30	2	3
161	Sheila	5	31	1	5

ID	Appraiser	Response	Sample	Trial	Standard
162	Sheila	5	31	2	5
163	Sheila	3	32	1	2
164	Sheila	3	32	2	2
165	Sheila	1	33	1	1
166	Sheila	1	33	2	1
167	Sheila	3	34	1	3
168	Sheila	3	34	2	3
169	Sheila	1	35	1	1
170	Sheila	1	35	2	1
171	Sheila	3	36	1	2
172	Sheila	3	36	2	2
173	Sheila	5	37	1	5
174	Sheila	5	37	2	5
175	Sheila	4	38	1	4
176	Sheila	4	38	2	4
177	Sheila	2	39	1	2
178	Sheila	2	39	2	2
179	Sheila	3	40	1	3
180	Sheila	3	40	2	3
181	Sheila	1	41	1	1
182	Sheila	1	41	2	1
183	Sheila	2	42	1	2
184	Sheila	2	42	2	2
185	Sheila	2	43	1	2
186	Sheila	2	43	2	2
187	Sheila	3	44	1	3
188	Sheila	3	44	2	3
189	Sheila	3	45	1	3
190	Sheila	3	45	2	3
191	Sheila	5	46	1	5
192	Sheila	5	46	2	5
193	Sheila	4	47	1	4
194	Sheila	4	47	2	4
195	Sheila	5	48	1	5
196	Sheila	5	48	2	5
197	Sheila	4	49	1	4
198	Sheila	4	49	2	4
199	Sheila	1	50	1	1
200	Sheila	1	50	2	1
201	Timothy	5	1	1	5
202	Timothy	5	1	2	5
203	Timothy	3	2	1	3
204	Timothy	4	2	2	3
205	Timothy	1	3	1	1
206	Timothy	1	3	2	1
207	Timothy	2	4	1	2

ID	Appraiser	Response	Sample	Trial	Standard
208	Timothy	2	4	2	2
209	Timothy	1	5	1	1
210	Timothy	1	5	2	1
211	Timothy	2	6	1	3
212	Timothy	2	6	2	3
213	Timothy	4	7	1	4
214	Timothy	4	7	2	4
215	Timothy	5	8	1	5
216	Timothy	5	8	2	5
217	Timothy	2	9	1	2
218	Timothy	2	9	2	2
219	Timothy	1	10	1	1
220	Timothy	1	10	2	1
221	Timothy	5	11	1	5
222	Timothy	5	11	2	5
223	Timothy	3	12	1	4
224	Timothy	4	12	2	4
225	Timothy	1	13	1	1
226	Timothy	1	13	2	1
227	Timothy	3	14	1	3
228	Timothy	3	14	2	3
229	Timothy	2	15	1	2
230	Timothy	2	15	2	2
231	Timothy	4	16	1	4
232	Timothy	4	16	2	4
233	Timothy	1	17	1	1
234	Timothy	1	17	2	1
235	Timothy	2	18	1	2
236	Timothy	2	18	2	2
237	Timothy	3	19	1	3
238	Timothy	3	19	2	3
239	Timothy	5	20	1	5
240	Timothy	5	20	2	5
241	Timothy	5	21	1	5
242	Timothy	5	21	2	5
243	Timothy	4	22	1	4
244	Timothy	4	22	2	4
245	Timothy	3	23	1	3
246	Timothy	3	23	2	3
247	Timothy	2	24	1	2
248	Timothy	2	24	2	2
249	Timothy	1	25	1	2
250	Timothy	2	25	2	2
251	Timothy	4	26	1	4
252	Timothy	4	26	2	4
253	Timothy	1	27	1	1

ID	Appraiser	Response	Sample	Trial	Standard
254	Timothy	1	27	2	1
255	Timothy	2	28	1	2
256	Timothy	2	28	2	2
257	Timothy	5	29	1	5
258	Timothy	5	29	2	5
259	Timothy	3	30	1	3
260	Timothy	3	30	2	3
261	Timothy	5	31	1	5
262	Timothy	5	31	2	5
263	Timothy	2	32	1	2
264	Timothy	1	32	2	2
265	Timothy	1	33	1	1
266	Timothy	1	33	2	1
267	Timothy	3	34	1	3
268	Timothy	4	34	2	3
269	Timothy	1	35	1	1
270	Timothy	1	35	2	1
271	Timothy	2	36	1	2
272	Timothy	2	36	2	2
273	Timothy	5	37	1	5
274	Timothy	5	37	2	5
275	Timothy	4	38	1	4
276	Timothy	4	38	2	4
277	Timothy	2	39	1	2
278	Timothy	2	39	2	2
279	Timothy	2	40	1	3
280	Timothy	2	40	2	3
281	Timothy	1	41	1	1
282	Timothy	1	41	2	1
283	Timothy	2	42	1	2
284	Timothy	2	42	2	2
285	Timothy	2	43	1	2
286	Timothy	1	43	2	2
287	Timothy	3	44	1	3
288	Timothy	3	44	2	3
289	Timothy	3	45	1	3
290	Timothy	3	45	2	3
291	Timothy	5	46	1	5
292	Timothy	5	46	2	5
293	Timothy	4	47	1	4
294	Timothy	4	47	2	4
295	Timothy	5	48	1	5
296	Timothy	5	48	2	5
297	Timothy	3	49	1	4
298	Timothy	4	49	2	4
299	Timothy	1	50	1	1

ID	Appraiser	Response	Sample	Trial	Standard
300	Timothy	1	50	2	1
301	Leslie	5	1	1	5
302	Leslie	5	1	2	5
303	Leslie	3	2	1	3
304	Leslie	3	2	2	3
305	Leslie	1	3	1	1
306	Leslie	1	3	2	1
307	Leslie	2	4	1	2
308	Leslie	2	4	2	2
309	Leslie	1	5	1	1
310	Leslie	1	5	2	1
311	Leslie	2	6	1	3
312	Leslie	2	6	2	3
313	Leslie	4	7	1	4
314	Leslie	4	7	2	4
315	Leslie	5	8	1	5
316	Leslie	5	8	2	5
317	Leslie	2	9	1	2
318	Leslie	2	9	2	2
319	Leslie	1	10	1	1
320	Leslie	1	10	2	1
321	Leslie	5	11	1	5
322	Leslie	5	11	2	5
323	Leslie	4	12	1	4
324	Leslie	4	12	2	4
325	Leslie	1	13	1	1
326	Leslie	1	13	2	1
327	Leslie	3	14	1	3
328	Leslie	3	14	2	3
329	Leslie	2	15	1	2
330	Leslie	2	15	2	2
331	Leslie	4	16	1	4
332	Leslie	4	16	2	4
333	Leslie	1	17	1	1
334	Leslie	1	17	2	1
335	Leslie	2	18	1	2
336	Leslie	3	18	2	2
337	Leslie	3	19	1	3
338	Leslie	3	19	2	3
339	Leslie	5	20	1	5
340	Leslie	5	20	2	5
341	Leslie	5	21	1	5
342	Leslie	5	21	2	5
343	Leslie	4	22	1	4
344	Leslie	4	22	2	4
345	Leslie	3	23	1	3

ID	Appraiser	Response	Sample	Trial	Standard
346	Leslie	3	23	2	3
347	Leslie	2	24	1	2
348	Leslie	2	24	2	2
349	Leslie	2	25	1	2
350	Leslie	2	25	2	2
351	Leslie	4	26	1	4
352	Leslie	4	26	2	4
353	Leslie	1	27	1	1
354	Leslie	1	27	2	1
355	Leslie	2	28	1	2
356	Leslie	2	28	2	2
357	Leslie	5	29	1	5
358	Leslie	5	29	2	5
359	Leslie	3	30	1	3
360	Leslie	3	30	2	3
361	Leslie	5	31	1	5
362	Leslie	5	31	2	5
363	Leslie	2	32	1	2
364	Leslie	3	32	2	2
365	Leslie	1	33	1	1
366	Leslie	1	33	2	1
367	Leslie	2	34	1	3
368	Leslie	2	34	2	3
369	Leslie	1	35	1	1
370	Leslie	1	35	2	1
371	Leslie	2	36	1	2
372	Leslie	2	36	2	2
373	Leslie	5	37	1	5
374	Leslie	5	37	2	5
375	Leslie	4	38	1	4
376	Leslie	4	38	2	4
377	Leslie	2	39	1	2
378	Leslie	2	39	2	2
379	Leslie	3	40	1	3
380	Leslie	3	40	2	3
381	Leslie	1	41	1	1
382	Leslie	1	41	2	1
383	Leslie	2	42	1	2
384	Leslie	2	42	2	2
385	Leslie	2	43	1	2
386	Leslie	2	43	2	2
387	Leslie	3	44	1	3
388	Leslie	3	44	2	3
389	Leslie	3	45	1	3
390	Leslie	3	45	2	3
391	Leslie	5	46	1	5

ID	Appraiser	Response	Sample	Trial	Standard
392	Leslie	5	46	2	5
393	Leslie	4	47	1	4
394	Leslie	4	47	2	4
395	Leslie	5	48	1	5
396	Leslie	5	48	2	5
397	Leslie	4	49	1	4
398	Leslie	4	49	2	4
399	Leslie	1	50	1	1
400	Leslie	1	50	2	1

Chi-square test of multiple proportions dataset

	Vehicle 1	Vehicle 2	Vehicle 3	Vehicle 4	Vehicle 5
Brake warrantee issues	200	20	50	50	0
No brake warrantee issues	749,800	649,980	249,950	49,950	450,000

Control chart—I-MR dataset

Brake caliper torsion
IMR

76.66
80.37
77.25
79.07
77.33
76.98
78.23
78.96
76.92
78.88
77.69
77.25
77.80
78.66
77.18
77.08
75.82
78.03
78.96
79.02
78.04
76.27
77.42
78.15
78.17
77.83

Brake caliper torsion
IMR
77.40
78.41
76.96
78.02

Control charts—P-chart dataset

Month	Cars produced	WarRec per 1000 PChart
Jan 2018	1013	192
Feb 2018	992	195
Mar 2018	960	191
Apr 2018	996	207
May 2018	983	194
Jun 2018	955	195
Jul 2018	951	195
Aug 2018	984	204
Sep 2018	1023	206
Oct 2018	986	209
Nov 2018	967	198
Dec 2018	1017	194
Jan 2019	1005	196
Feb 2019	969	195
Mar 2019	1022	203
Apr 2019	969	208
May 2019	970	197
Jun 2019	971	207
Jul 2019	1014	209
Aug 2019	954	205
Sep 2019	1007	195
Oct 2019	1008	209
Nov 2019	964	200
Dec 2019	1000	200
Jan 2020	1016	195
Feb 2020	1017	199
Mar 2020	991	196
Apr 2020	1008	206
May 2020	1020	195
Jun 2020	968	206
Jul 2020	970	197
Aug 2020	971	207
Sep 2020	1014	209
Oct 2020	954	205
Nov 2020	1007	195
Dec 2020	1008	209

Control chart—Xbar-R dataset

Subgroup	XBarR caliper torsion
1	76
1	77
1	78
2	77
2	80
2	79
3	80
3	77
3	80
4	80
4	76
4	80
5	76
5	85
5	78
6	75
6	80
6	76
7	78
7	78
7	78
8	78
8	75
8	75
9	77
9	78
9	80
10	80
10	62
10	75
11	76
11	79
11	77
12	75
12	79
12	80
13	75
13	75
13	79
14	78
14	77
14	78
15	76
15	76
15	80

Subgroup	XBarR caliper torsion
16	75
16	79
16	80
17	79
17	80
17	80
18	76
18	80
18	75
19	75
19	78
19	79
20	76
20	76
20	76
21	80
21	76
21	77
22	77
22	76
22	77
23	76
23	75
23	80
24	78
24	76
24	79

Control chart—U-chart dataset

Month_1	Airbags produced	Defects identified	Shift
Jan 2018	1013	192	Before
Feb 2018	992	195	Before
Mar 2018	960	100	Before
Apr 2018	996	207	Before
May 2018	983	194	Before
Jun 2018	955	195	Before
Jul 2018	951	195	Before
Aug 2018	984	204	Before
Sep 2018	1023	206	Before
Oct 2018	986	209	Before
Nov 2018	967	198	Before
Dec 2018	1017	194	Before
Jan 2019	900	196	Before
Feb 2019	969	195	Before
Mar 2019	1022	203	Before

Month_1	Airbags produced	Defects identified	Shift
Apr 2019	969	208	Before
May 2019	970	197	Before
Jun 2019	971	207	Before
Jul 2019	1014	209	Before
Aug 2019	954	205	Before
Sep 2019	1007	195	Before
Oct 2019	1200	209	Before
Nov 2019	964	200	Before
Dec 2019	1000	200	Before
Jan 2020	1016	195	Before
Feb 2020	1017	199	Before
Mar 2020	991	196	Before
Apr 2020	1008	206	Before
May 2020	1020	195	Before
Jun 2020	968	350	After
Jul 2020	970	325	After
Aug 2020	971	400	After
Sep 2020	1014	402	After
Oct 2020	954	350	After
Nov 2020	1007	325	After
Dec 2020	1008	401	After

Run chart dataset

Ex caliper mold diameter IMR	Ex caliper mold diameter RunCh
19.50	19.50
20.90	20.90
21.20	21.20
21.36	21.36
22.22	22.22
23.41	23.41
24.10	24.10
24.80	24.80
25.20	25.20
25.80	25.80
23.31	
24.19	
23.33	
23.30	
24.23	
23.67	
22.75	
23.33	
23.66	
23.36	
23.15	

Ex caliper mold diameter IMR	Ex caliper mold diameter RunCh
23.20	
23.56	
21.40	
22.61	
23.54	

DOE dataset

StdOrder	RunOrder	CenterPt	Blocks	Temp	Vacuum	Time	Glue	Airbag seal strength	Airbag seal strength with interaction
2	1	1	1	200	30	100	Brand A	1090	1390
19	2	1	1	180	50	100	Brand A	929	1379
15	3	1	1	180	50	200	Brand B	1010	1460
20	4	1	1	200	50	100	Brand A	1027	1527
32	5	1	1	200	50	200	Brand B	1122	1622
5	6	1	1	180	30	200	Brand A	1070	1340
4	7	1	1	200	50	100	Brand A	1030	1530
21	8	1	1	180	30	200	Brand A	1081	1351
17	9	1	1	180	30	100	Brand A	987	1257
31	10	1	1	180	50	200	Brand B	1024	1474
24	11	1	1	200	50	200	Brand A	1122	1622
1	12	1	1	180	30	100	Brand A	990	1260
22	13	1	1	200	30	200	Brand A	1181	1481
6	14	1	1	200	30	200	Brand A	1170	1470
29	15	1	1	180	30	200	Brand B	1082	1352
14	16	1	1	200	30	200	Brand B	1170	1470

StdOrder	RunOrder	CenterPt	Blocks	Temp	Vacuum	Time	Glue	Airbag seal strength	Airbag seal strength with interaction
12	17	1	1	200	50	100	Brand B	1030	1530
10	18	1	1	200	30	100	Brand B	1090	1390
18	19	1	1	200	30	100	Brand A	1085	1385
30	20	1	1	200	30	200	Brand B	1181	1481
11	21	1	1	180	50	100	Brand B	930	1380
28	22	1	1	200	50	100	Brand B	1026	1526
25	23	1	1	180	30	100	Brand B	988	1258
27	24	1	1	180	50	100	Brand B	928	1378
16	25	1	1	200	50	200	Brand B	1110	1610
13	26	1	1	180	30	200	Brand B	1070	1340
9	27	1	1	180	30	100	Brand B	990	1260
23	28	1	1	180	50	200	Brand A	1024	1474
3	29	1	1	180	50	100	Brand A	930	1380
8	30	1	1	200	50	200	Brand A	1110	1610
7	31	1	1	180	50	200	Brand A	1010	1460
26	32	1	1	200	30	100	Brand B	1085	1385

Gage R&R dataset

Part	Assessor	Response
1	A	82.29
1	A	82.41
1	A	82.64
2	A	81.44
2	A	81.32
2	A	81.42
3	A	83.34
3	A	83.17
3	A	83.27

Part	Assessor	Response
4	A	82.47
4	A	82.5
4	A	82.64
5	A	81.2
5	A	81.08
5	A	81.16
6	A	82.02
6	A	81.89
6	A	81.79
7	A	82.59
7	A	82.75
7	A	82.66
8	A	81.69
8	A	81.8
8	A	81.83
9	A	84.26
9	A	83.99
9	A	84.01
10	A	80.64
10	A	80.75
10	A	80.69
1	B	82.08
1	B	82.25
1	B	82.07
2	B	81.53
2	B	80.78
2	B	81.32
3	B	83.19
3	B	82.94
3	B	83.34
4	B	82.01
4	B	83.03
4	B	82.2
5	B	81.44
5	B	80.8
5	B	80.72
6	B	81.8
6	B	82.22
6	B	82.06
7	B	82.47
7	B	82.55
7	B	82.83
8	B	81.37
8	B	82.08
8	B	81.66
9	B	83.8

Part	Assessor	Response
9	B	84.12
9	B	84.19
10	B	80.32
10	B	80.38
10	B	80.5
1	C	82.04
1	C	81.89
1	C	81.85
2	C	80.62
2	C	80.87
2	C	81.04
3	C	82.88
3	C	83.09
3	C	82.67
4	C	82.14
4	C	82.2
4	C	82.11
5	C	80.54
5	C	80.93
5	C	80.55
6	C	81.71
6	C	81.33
6	C	81.51
7	C	82.02
7	C	82.01
7	C	82.21
8	C	81.54
8	C	81.44
8	C	81.51
9	C	83.77
9	C	83.45
9	C	83.87
10	C	80.51
10	C	80.23
10	C	79.84

Paired T test dataset

Sample	Device A	Device B	Differences
1	0.030837	0.033921	0.003083709
2	0.02889	0.031779	0.002888979
3	0.030107	0.033118	0.003010699
4	0.029378	0.032316	0.00293784
5	0.029524	0.032477	0.002952444
6	0.031008	0.034109	0.003100792
7	0.031211	0.034332	0.003121102

Sample	Device A	Device B	Differences
8	0.029385	0.032324	0.002938535
9	0.030467	0.033514	0.003046686
10	0.031654	0.03482	0.003165437

Pareto summarized data

Category	Count of defect
Airbags	344
Brakes	487
Defect	20
Electronics	60
Emissions	80
Hardware	20
Propulsion	40

Process capability datasets

Brake caliper torsion IMR	Airbag tensile strength
76.66	40.86
80.37	13.82
77.25	3.92
79.07	4.06
77.33	5.89
76.98	5.05
78.23	4.95
78.96	4.37
76.92	53.33
78.88	8.35
77.69	18.38
77.25	11.87
77.80	7.29
78.66	2.07
77.18	16.04
77.08	7.14
75.82	20.22
78.03	2.69
78.96	0.63
79.02	12.88
78.04	1.39
76.27	17.97
77.42	8.96
78.15	1.70
78.17	3.63
77.83	6.84
77.40	19.11
78.41	0.42
76.96	5.48
78.02	11.53

Single regression dataset

Ex caliper mold diameter	Ex deslagger abrasion setting
19.50	57.00
20.90	61.00
21.20	62.00
21.36	62.10
22.22	65.00
23.41	68.00
24.10	70.00
24.80	72.00
25.20	74.00
25.80	75.00
23.31	68.00
24.19	71.00
23.33	68.15
23.30	68.00
24.23	71.00
23.67	69.00
22.75	66.00
23.33	68.00
23.66	69.00
23.36	68.00
23.15	68.05
23.20	68.10
23.56	69.00
21.40	63.00
22.61	66.00
23.54	69.00

Multiple regression dataset

MR ex caliper mold diameter_1	Ex deslagger abrasion setting_1	Ambient temperature	Relative humidity	Lub pH	Abraizer machine
19.50	57	76.82	80.65	3.5	Machine A
20.90	61	77.80	67.47	3.6	Machine A
21.20	62	69.10	37.14	3.5	Machine A
21.36	62.1	74.89	74.24	3.7	Machine A
22.22	65	72.38	65.58	3.8	Machine A
23.41	68	74.09	45.76	4.01	Machine B
24.10	70	84.26	38.18	4.2	Machine B
24.80	72	75.65	79.97	4.3	Machine B
25.20	74	78.69	46.81	4.25	Machine B
25.80	75	78.47	59.45	4.45	Machine B
23.31	68	79.52	41.17	4	Machine B
24.19	71	72.91	69.91	4.15	Machine B
23.33	68.15	75.60	83.67	3.99	Machine B

MR ex caliper mold diameter_1	Ex deslagger abrasion setting_1	Ambient temperature	Relative humidity	Lub pH	Abraizer machine
23.30	68	76.46	83.50	3.5	Machine B
24.23	71	73.49	70.05	4.2	Machine B
23.67	69	84.33	50.61	4.1	Machine B
22.75	66	69.59	50.89	3.9	Machine A
23.33	68	79.69	63.50	4	Machine B
23.66	69	74.67	39.32	4.1	Machine B
23.36	68	70.06	62.19	3.85	Machine B
23.15	68.05	83.40	30.29	3.99	Machine B
23.20	68.1	78.27	64.58	4.01	Machine B
23.56	69	78.37	71.50	4	Machine B
21.40	63	72.49	39.23	3.59	Machine A
22.61	66	70.29	76.34	3.75	Machine A
23.54	69	73.11	39.37	4.02	Machine B

Test of two proportions dataset

Category	Vendor 1	Vendor 2
Airbags with issues	200	180
Total airbags used	50,000	40,000
Percent defective	0.400%	0.450%

Two sample T test dataset

Month_1	Airbags produced	Defects identified	Defects per airbag	Shift
Jan 2018	1013	192	19.0%	Before
Feb 2018	992	195	19.7%	Before
Mar 2018	960	100	10.4%	Before
Apr 2018	996	207	20.8%	Before
May 2018	983	194	19.7%	Before
Jun 2018	955	195	20.4%	Before
Jul 2018	951	195	20.5%	Before
Aug 2018	984	204	20.7%	Before
Sep 2018	1023	206	20.1%	Before
Oct 2018	986	209	21.2%	Before
Nov 2018	967	198	20.5%	Before
Dec 2018	1017	194	19.1%	Before
Jan 2019	900	196	21.8%	Before
Feb 2019	969	195	20.1%	Before
Mar 2019	1022	203	19.9%	Before
Apr 2019	969	208	21.5%	Before
May 2019	970	197	20.3%	Before
Jun 2019	971	207	21.3%	Before
Jul 2019	1014	209	20.6%	Before
Aug 2019	954	205	21.5%	Before
Sep 2019	1007	195	19.4%	Before

Month_1	Airbags produced	Defects identified	Defects per airbag	Shift
Oct 2019	1200	209	17.4%	Before
Nov 2019	964	200	20.7%	Before
Dec 2019	1000	200	20.0%	Before
Jan 2020	1016	195	19.2%	Before
Feb 2020	1017	199	19.6%	Before
Mar 2020	991	196	19.8%	Before
Apr 2020	1008	206	20.4%	Before
May 2020	1020	195	19.1%	Before
Jun 2020	968	350	36.2%	After
Jul 2020	970	325	33.5%	After
Aug 2020	971	400	41.2%	After
Sep 2020	1014	402	39.6%	After
Oct 2020	954	350	36.7%	After
Nov 2020	1007	325	32.3%	After
Dec 2020	1008	401	39.8%	After
Jan 2021	954	350	36.7%	After
Feb 2021	970	325	33.5%	After
Mar 2021	971	400	41.2%	After
Apr 2021	1014	402	39.6%	After
May 2021	954	350	36.7%	After
Jun 2021	1007	325	32.3%	After
Jul 2021	954	350	36.7%	After
Aug 2021	1014	402	39.6%	After
Sep 2021	954	350	36.7%	After

Epilogue

Albert was quite pleased. Three years had passed since Charlie first called him and made an offer he couldn't refuse. Since then, the warrantee work and recalls had reduced dramatically. Sales were back up, and everyone was making lots of money. Shareholders had realized a 50% increase in their share price.

Charlie had started to change the culture in KIND Karz and at their primary partners. Each location had people certified in Green and Black Belt Six Sigma and had formed rapid response teams that included people from the shop floor to management with demonstrable results.

Albert had developed in-house two talented Master Black Belts, Chris and Bekah, and they were really running things now. This was how he had planned it. Now was time to take early retirement and maybe do something else after a few months of doing nothing. At 58, he really didn't need to make any more money. Also, Charlie had really come through, and in addition to a very nice salary along with the increase in the stock market, he had a couple million dollars' worth of stock and options just waiting to be spent. It was time to take his mother, Ana, on that much needed European cross-country tour. He might even let his wife come along too.

As he walked towards Charlie's new office, he stopped, took a deep breath, and smiled. It would honestly be good to just be friends with Charlie versus working for him. Being friends was easier, and more fun he had found. But he had kept his word—3 years—and had really led the company through some significant challenges.

Charlie kept his chief of staff, Jonathan, as kind of a guard with an office in front of his. While he had an open-door policy in principle, Jonathan was quick to stop someone if Charlie was in one of his grumpy *alone times*. Albert was an exception. As he walked past Jonathan's office, he looked out and said, "Watch out – he's in one of those moods!"

And there Charlie sat—head in his hands, looking at paper spread over his desk. He still preferred paper over a screen. "Come in, come in," he said without looking up. "I've really need to talk to you."

"Actually," Albert said, "I have something I want to …".

"You won't believe this," interrupted Charlie. "It's that airbag vendor again. Now they're asking for more advanced notice on orders – six months before I need them!

© The Author(s), under exclusive license to Springer Nature Switzerland AG 2022
T. D. Blackburn, *Six Sigma*, https://doi.org/10.1007/978-3-030-96213-5

And when we get there, they don't have the full order quantity ready for us to pick up. I know we're ordering more now that your Six Sigma thing has turned things around for us, but this is ridiculous. It doesn't help that we're twice the quantity per car as the big auto guys, but I guess that's something else that sets us apart. But I digress – every year the unit cost is increasing higher than inflation. Get your Six Sigma crew on this ASAP."

"Well, that's actually Lean, not Six sigma. More like the TPS stuff you like. But I came here to give you my …".

"Stop, stop." Charlie waived his hands and shook his head. "I just need another year – that's all. Just one more year. You say that's Lean? Ok, make it happen. Set it up. I don't care how. I'll make it worth your while …".

References

A.I.A.G. Measurement systems analysis reference manual. (1990).

Antony, J., Snee, R., & Hoerl, R. (2017). Lean six sigma: Yesterday, today, and tomorrow. *The International Journal of Quality & Reliability Management*, 1073–1093.

Bothe, D. R. (1997). *Measuring process capability*. McGraw Hill.

Box, G., & Draper, N. (1987). *Empirical model-building and response surfaces*. Wiley.

Box, G., Hunter, W. G., & Hunter, J. (1978). *Statistics for experimenters*. Wiley.

Cintas, P., Almagro, L., & Labres, X.-M. (2012). Industrial statistics with minitab.

Dorfman, R. (1981). The meaning of internal rates of return. *The Journal of Finance*, 1011–1021.

Federico, M., & Beaty, R. (2003). *Six sigma team pocket guide*. McGraw-Hill.

George, M. L., Rowlands, D., Price, M., & Maxey, J. (2005). *Lean six sigma pocket toolbook*. McGraw-Hill.

Gupta, H. C., Guttman, I., & Jayalath, K. P. (2020). *Statistics and probability with applications for engineers and scientists using MINITAB, R and JMP* (2nd ed.). Wiley.

(n.d.). https://www.businessballs.com/self-management/paretos-80-20-rule-theory/.

Joiner, B. L. (1980). Lurking variables: Some examples. *The American Statistician*, 227–233.

Liker, J. K. (2020). *The toyota way* (2nd ed.). McGraw Hill.

Lochner, R. H., & Matar, J. E. (1990). *Designing for quality: An introduction to the best of Taguchi and Western methods of statistical experimental design*. Quality Resources.

Minitab Online Help. (n.d.). https://www.minitab.com/en-us/support/.

Moen, R. D., Nolan, T. W., & Provost, L. P. (1989). *Planned experimentation to improve quality*. Associates in Process Improvement.

Myers, R. (1976). *Response surface methodology*. Virginia Polytechnic Institute and State University.

Patterson, G., & Fedrico, M. (2006). *Six sigma champions pocket guide*. Rath & Strong.

Ryan, B., Joiner, B. L., & Cryer, J. (2012). *Minitab handbook*. Cengage.

Sleeper, A. (2012). *Minitab deMystified*. McGraw Hill.

Snee, R. D., Hare, L. B., & Trout, J. R. (1985). *Experiments in industry: Design, analysis, and interpretation*. American Society for Quality Control.

Stagliano, A. A. (2004). *Six sigma advanced tools pocket guide*. McGraw Hill.

Wheeler, D. J. (1987). *Understanding industrial experimentation*. SPC Press, Inc.

Wheeler, D. J., & Lyday, R. W. (1989). *Evaluating the measurement process*. SPC Press Inc.

© The Author(s), under exclusive license to Springer Nature Switzerland AG 2022
T. D. Blackburn, *Six Sigma*, https://doi.org/10.1007/978-3-030-96213-5

Index

© The Author(s), under exclusive license to Springer Nature Switzerland AG 2022
T. D. Blackburn, *Six Sigma*, https://doi.org/10.1007/978-3-030-96213-5

Printed in the United States
by Baker & Taylor Publisher Services